High School Probability: A New Perspective

For Any High School Curriculum, Including IB, A-Level, AP, and More

Spring Seeds

Published by Spring Seeds

Print ISBN: 979-8-89571-041-8

Contents

Preface

Probability theory is the branch of mathematics that studies and reveals random phenomena. It is also a core course to cultivate students' careful and rigorous thinking and improve their mathematical skills. In the teaching of probability theory, the study of probability concepts and principles is of special importance. It is a necessary way to understand and master probability knowledge.

We have read and studied many textbooks on probability. We found that these textbooks are not very specific and simple in their elaboration and explanation of the concepts and principles of probability. Therefore, we have developed this textbook in order to meet the needs of students to learn probability.

This textbook has been used in the teaching of IB higher level mathematics course and has been highly praised by students. Students find the concepts and principles of this textbook particularly clear, concrete and simple, and logic is strong. Therefore, students find it easy to learn and master probability knowledge. Moreover, students have also achieved very good results in the study of probability.

It can be said that this textbook has completely ended the long-term difficult state of teaching probability theory. It makes the study of probability theory very easy. This has been verified in the long term teaching practice.

To make this textbook fully adaptable, it contains essential contents applicable to different high school curricula, as well as contents marked with * that can be selected by various high schools.

Spring Seeds

2024

This page is intentionally left blank.

Chapter 1 Permutation and Combination

Knowledge of permutation and combination is a computational basis for probability theory. Permutation and combination study the situations where r elements are chosen from n elements ($r \le n$) to form ordered permutations or unordered combinations.

1.1. Permutation

To show clearly the concept of permutation, let us read two examples first.

Example 1. For one meeting, choose a chairperson and a secretary from four persons: a, b, c, and d. How many different ways to choose (that is, what is the number of permutations)?

Solution. There are four choices for the chairperson. After this, there are three choices for the secretary from the remaining three persons.

Hence number of permutations $\quad {}^nP_r = {}^4P_2 = 4\cdot(4-1) = 12$

The symbol ${}^nP_r$ means the number of permutations when r elements are chosen randomly from n different elements ($r \le n$) to form ordered permutations.

Here, we list all possible outcomes of the ordered permutations in this example (represented by a set).

$$\begin{aligned} A &= \{(x_1, x_2)\} \\ &= \{(a, b), (a, c), (a, d), (b, a), (b, c), (b, d), (c, a), (c, b), (c, d), (d, a), (d, b), (d, c)\} \end{aligned}$$

Element $(x_1, x_2, ..., x_n)$ in such a set is called *n*-variables ordered-number-group.

Except $x_1 = x_2 = ... = x_n$

$$(x_1, x_2, ..., x_n) \neq (x_2, x_1, ..., x_n) \neq ... \neq (x_n, ..., x_2, x_1)$$

Therefore, in this example (a, b) ≠ (b, a),…, (c, d) ≠ (d, c).

The number of elements in set A is *n*(A) = 12.

So $\quad {}^4P_2 = n(\text{A}) = 4 \cdot (4 - 1) = 12$

Example 2. Set A has five elements, where A = {1, 2, 3, 4, 5}. Calculate

a) the number of permutations when 3 elements are chosen from set A to form ordered permutations.
b) the number of permutations when 5 elements are chosen from set A to form ordered permutations.

Solution.

a) We know $n = n(\text{A}) = 5$, $r = 3$.

So $\quad {}^nP_r = {}^5P_3 = 5 \cdot (5 - 1)(5 - 2) = 5 \cdot 4 \cdot 3 = 60$

b) We know $n = n(\text{A}) = 5$, $r = 5$.

So $\quad {}^nP_r = {}^5P_5 = 5 \cdot (5 - 1)(5 - 2)(5 - 3)(5 - 4) = 5 \cdot 4 \cdot 3 \cdot 2 \cdot 1 = 120$

Here ${}^5P_5$ is called the complete permutation or 5-factorial, denoted as ${}^5P_5 = 5!$.

That is, $\quad 5! = 5 \cdot 4 \cdot 3 \cdot 2 \cdot 1$

In the same way, we can calculate *n*-factorial. $\quad n! = {}^nP_n$

n-factorial formula $\quad n! = n(n - 1)(n - 2) \cdots 3 \cdot 2 \cdot 1 \quad (n \in Z^+)$

$0! = 1$

By deriving from the above examples, we can get the formula for calculating the number of permutations.

$$ {}^{n}P_{r} = \frac{n!}{(n-r)!} = \underbrace{n\cdot(n-1)(n-2)\cdots(n-(r-1))}_{r\ \ factors} \qquad (0 \le r \le n) $$

1.2. Combination

Below, we illustrate the concept of combination with two examples.

Example 1. There are 5 elements in set A, A = {1, 2, 3, 4, 5}. Evaluate the number of combinations when 3 elements are chosen from set A to form unordered combinations.

Solution. We know $n = n(\mathrm{A}) = 5$, $r = 3$.

So $$\text{number of combinations} = \binom{n}{r} = \binom{5}{3} = \frac{5\times4\times3}{3\times2\times1} = 10$$

Symbol $\binom{n}{r}$ means the number of combinations when r elements are chosen from n different elements ($r \le n$) to form unordered combinations.

Here, we list all possible outcomes of unordered combinations in this example (represented by a set).

$$\begin{aligned} \mathrm{B} &= \{\ \{x_1, x_2, x_3\}\ \} \\ &= \{\ \{1, 2, 3\}, \{1, 2, 4\}, \{1, 2, 5\}, \{1, 3, 4\}, \{1, 3, 5\}, \{1, 4, 5\}, \{2, 3, 4\}, \\ &\quad \{2, 3, 5\}, \{2, 4, 5\}, \{3, 4, 5\}\} \end{aligned}$$

For unordered combinations, it is always true that

$$\{x_1, x_2, \ldots, x_n\} = \{x_2, x_1, \ldots, x_n\} = \ldots = \{x_n, \ldots, x_2, x_1\}$$

So, in this example

$$\{x_1, x_2, x_3\} = \{x_2, x_1, x_3\} = \ldots = \{x_3, x_2, x_1\}$$

For example $\{1, 2, 3\} = \{1, 3, 2\} = \{2, 1, 3\} = \{2, 3, 1\} = \{3, 1, 2\} = \{3, 2, 1\}$

Hence, the number of elements in set B is $n(\text{B}) = 10$.

So $$\binom{5}{3} = n(\text{B}) = \frac{5\times4\times3}{3\times2\times1} = 10$$

Example 2. 5 cards are drawn from a pack of 52 playing cards. How many possible drawings are there (that is, what is the number of combinations)?

Solution. It is given that $n = 52$, $r = 5$.

So $$\binom{n}{r} = \binom{52}{5} = \frac{52\times51\times50\times49\times48}{5\times4\times3\times2\times1} = 2598960$$

Deriving from the above examples, we can get the formula for the number of combinations.

$$\binom{n}{r} = \frac{\overbrace{n\cdot(n-1)(n-2)\cdots(n-(r-1))}^{r\ \ factors}}{r!} = \frac{n!}{r!(n-r)!} \qquad 0 \le r \le n$$

When $r = 0$, $\binom{n}{0} = 1$.

When $r = n$, $\binom{n}{n} = 1$.

1.3. Examples

Example 1. 5 boys and 4 girls sit in a row. How many ways can this be arranged

a. if there is no restriction?

b. if the boys and girls each sit together?

Solution.

a. Since there is no restriction, we will put 9 persons in permutation. Thus, we have

$$ {}^{9}P_{9} = 9! = 362880 \quad \text{ways} $$

b. If the boys and girls each sit together, we need to first treat all boys as one big boy and all girls as one big girl, then there are 2! ways to arrange them. Then there are 5! ways to arrange all the boys and 4! ways to arrange all the girls. Thus, altogether, there are

$$ 2!5!4! = 5760 \quad \text{ways} $$

Example 2. How many different ways to arrange all the letters from the word COMMITTEE?

Solution. There are altogether 9 letters in this word, so putting all 9 letters in permutation, we have ${}^{9}P_{9} = 9! = 362880$ ways. However, there are two letters M, which will give the same permutations. We need to exclude this so $\frac{9!}{2!}$. We also have two letters T and two letters E, thus altogether, we have

$$ \frac{9!}{2!2!2!} = 45360 \qquad \text{different ways} $$

Example 3. 6 different colour flags can be hung in a vertical line to form different signals.

a. How many different signals can be formed if all 6 colour flags are used?

b. How many different signals can be formed if any number of colour flags is used?

Solution. a. If all 6 colour flags are used, we will put all 6 colours in permutation so we have

$$ {}^{6}P_{6} = 6! = 720 \quad \text{different signals} $$

b. If any number of colour flags can be used, we can choose 1 colour to form a signal, and then we have ${}^{6}P_{1}$ signals. Or we can choose 2 colours to form a signal, and then we have ${}^{6}P_{2}$ signals. Similarly, if 3, 4, 5 or 6 colours are chosen, we will have ${}^{6}P_{3}$, ${}^{6}P_{4}$, ${}^{6}P_{5}$ and ${}^{6}P_{6}$ signals respectively. Thus, altogether, we have

$$ {}^{6}P_{1} + {}^{6}P_{2} + {}^{6}P_{3} + {}^{6}P_{4} + {}^{6}P_{5} + {}^{6}P_{6} = 1956 \qquad \text{different signals} $$

Example 4.

a. How many ways can a committee of 3 men and 2 women be chosen from 7 men and 5 women?
b. Among the 12 people, a married couple just divorced and did not want to be chosen together. How many ways can a committee be formed?

Solution. a. We need to choose 3 men from 7 men, which can be done in $\binom{7}{3}$ ways.

We need to choose 2 women from 5 women, which can be done in $\binom{5}{2}$ ways. Thus,

altogether, there are

$$ \binom{7}{3}\binom{5}{2} = 350 \quad \text{ways} $$

b. Method 1 There are three cases in which the couple is not chosen simultaneously.

Case 1. The couple are both not chosen, so we need to choose 3 men from the remaining 6 men and 2 women from the remaining 4 women, and there are $\binom{6}{3}\binom{4}{2}$ ways.

Case 2. The husband is chosen, and the wife is not chosen. We need to choose 2 men from the remaining 6 men and 2 women from the remaining 4 women. Thus, we have $\binom{6}{2}\binom{4}{2}$ ways.

Case 3. The husband is not chosen, but the wife is chosen. We need to choose 3 men from the remaining 6 men and 1 woman from the remaining 4 women. Thus, we have $\binom{6}{3}\binom{4}{1}$ ways.

Adding all these three cases together, we have

$$\binom{6}{3}\binom{4}{2}+\binom{6}{2}\binom{4}{2}+\binom{6}{3}\binom{4}{1}=290 \quad \text{ways}$$

Method 2. If the couple is both chosen, we need to choose 2 men from the remaining 6 men and 1 woman from the remaining 4 women. Thus, we have $\binom{6}{2}\binom{4}{1}$ ways.

Subtracting this from the total ways with no restriction (we get in a.) will give us the number of ways that the couple will not be both chosen.

$$\binom{7}{3}\binom{5}{2}-\binom{6}{2}\binom{4}{1}=290$$

Example 5. There are 11 boys and 20 girls in a debate club. The club must choose a team of 4 persons for a debate conference. How many different conference teams are possible

a. if the team consists of 2 boys and 2 girls?

b. if the team consists of more girls than boys?

c. if the team consists of people of the same sex?

Solution. a. If the team consists of 2 boys and 2 girls, there are $\binom{11}{2}$ ways to choose 2 boys from 11 boys. And there are $\binom{20}{2}$ ways to choose 2 girls from 20. Thus, altogether, there are

$$\binom{11}{2}\binom{20}{2} = 10450 \qquad \text{teams}$$

b. If the team has more girls than boys, it could be 3 girls and 1 boy or 4 girls only. So, there are altogether

$$\binom{20}{3}\binom{11}{1} + \binom{20}{4} = 17385 \quad \text{teams}$$

c. If the team consists of the same sex, then it could be either 4 boys only or 4 girls only. Altogether, we have

$$\binom{11}{4} + \binom{20}{4} = 5175 \qquad \text{teams}$$

Exercise 1

1. A hotel director has to select a manager, an assistant manager and a restaurant manager from a list of 12 people. How many ways can this be done?

2. In one country, license plates have three letters from a 26-letter alphabet followed by three digits from 0 to 9.

- **a.** How many license plates can be created if there are no other restrictions?
- **b.** What if only letters cannot be repeated?
- **c.** What if only digits cannot be repeated?
- **d.** What if both letters and digits cannot be repeated?

3. At a dog show, the judge must choose first-, second- and third-place finishers from a group of 28 dogs. How many ways can this be done?

4. A class of 30 students is going on a field trip. The first bus has only 15 seats, so the teacher has to choose 15 students to take the first bus. The remaining students and the teacher wait for the second bus.

- **a.** If there is no restriction, how many ways to choose 15 students?
- **b.** If three students are friends and they want to travel on the same bus, how many ways to choose 15 students?

5. A teacher starts each lesson by choosing 3 students to present solutions to homework problems to the class. If there are 30 students in the class

- **a.** how many ways can the teacher make the selection?
- **b.** Anton and Maya presented during the last lesson. The teacher wants to select other students. Then, how many ways can the teacher make the selection?

6. The jury consists of 7 persons selected from 10 male and 10 female candidates.

- **a.** How many different ways of selection are there?
- **b.** Out of the 20 candidates, two are from the same family and cannot be selected simultaneously. How many different ways of selection are there?
- **c.** If the jury must have at least 3 persons from each sex, how many different ways of selection are there?

Chapter 2 Basic concepts in probability theory

2.1. Random experiment, Sample space and Classical probability

A wide variety of phenomena exists in the natural and social world. One type of phenomenon will definitely occur under certain conditions. For instance, if a stone is thrown upwards, it will definitely fall down. This phenomenon is called a **deterministic phenomenon.** Another type of phenomenon still exists in the natural and social world. For example, throw a coin upwards under the same conditions. When the coin comes down, the coin most likely gives a head or a tail. Every time before we throw the coin, we never know for sure what the outcome will be. This type of phenomenon possesses the following characteristics: under certain conditions, it can produce this or that outcome. A definite outcome cannot be known before an experiment or observation. However, after long-term practice and research, people have found that in a large number of repeated experiments or observations, the outcomes show a certain regularity. For instance, when we repeatedly toss a coin many times, the number of heads will be approximately half the total number of tosses. This kind of regularity, which manifests in a large number of repeated experiments or observations, is statistical regularity.

This type of phenomenon, which shows uncertainty in the results of individual experiments but has a certain regularity in a large number of repeated experiments, is called a **random phenomenon.** Probability theory and statistics are branches of mathematics that study and reveal the statistical regularity of random phenomena.

Random experiment

An experiment with the following three characteristics is a random experiment:

1. It can be repeated under the same conditions.
2. Every experiment has more than one possible outcome. Furthermore, all possible outcomes of the experiment can be known in advance.
3. Before each experiment, it is impossible to determine which outcome will occur.

In probability theory, we study random events by studying random experiments.

Sample space

Although the outcome of each random experiment S is not known in advance, the sample space U, which consists of all possible outcomes of the experiment, is known. The elements of the sample space U are called the possible outcomes of random experiment S.

In the following, we will read some examples of random experiments.

S_1: Toss a coin once. Observe the outcome of the head and tail.

$U_1 = \{\text{head, tail}\} = \{H, T\}$ $n(U_1) = 2$

S_2: Toss a coin three times. Observe the outcome of the head and tail.

$U_2 = \{HHH, HHT, HTH, HTT, THH, THT, TTH, TTT\}$ $n(U_2) = 8$

S_3: Toss a coin three times. Observe the number of times of the occurrence of head.

$U_3 = \{0, 1, 2, 3\}$ $n(U_3) = 4$

S_4: Throw a die once. Observe the number obtained.

$U_4 = \{1, 2, 3, 4, 5, 6\}$ $n(U_4) = 6$

Note:

The purpose of the experiment determines the elements of the sample space. For instance, experiments S_2 and S_3 are the same: tossing a coin three times. Due to the different experiment purposes, the sample space U_2 and U_3 are thus different.

Random event

When conducting a random experiment, people are only interested in possible outcomes that satisfy certain conditions. For example, we observe the even number obtained when throwing a die. In S_4, we are concerned with subset A, which consists of possible outcomes that satisfy the condition "even number", $A = \{2, 4, 6\}$. We call A a random event of random experiment S_4.

Generally, we refer to the subset of the sample space of the experiment S as the random event of S or event for short. In each experiment, if and only if one possible outcome of this subset occurs, we say this event occurs.

In particular, a single-element set consisting of a single outcome is called a simple event. Experiment S_1 has two simple events, {H} and {T}. Experiment S_4 has six simple events, {1}, {2},…, {6}.

The sample space U includes all possible outcomes. U is a subset of U itself. It always happens in every experiment. So $P(U) = 1$, and it is called a definite event. The empty set ∅ does not include any possible outcome. It is also a subset of the sample space. It does not happen in every experiment. It is called an impossible event. $P(\varnothing) = 0$.

Relative frequency and probability

Except for definite events and impossible events, an event may or may not occur in an experiment. We often want to know how likely a particular event will occur in an experiment and find a suitable number to represent the possibility of a certain event occurring in an experiment. Therefore, relative frequency was first introduced, which describes how often a certain event occurs. From this, probability was introduced, which represents the likelihood of an event occurring in an experiment.

Definition of relative frequency

Conduct an experiment n times under the same condition. Among n trials, event A occurs f_A times and f_A is called the occurrence frequency of event A. The ratio $\frac{f_A}{n}$ is called the relative frequency of the occurrence of event A, denoted by $f_n(A)$. That is,

relative frequency $\qquad f_n(A)=\frac{f_A}{n}$

it is easy to know $\qquad 0 \le f_n(A) \le 1$

Example. Consider the experiment "tossing a coin". We toss a coin 5 times, 50 times and 500 times. Each of these experiments is repeated 10 times. The data from the experiments are shown in Table 2.1, where f_H represents the frequency at which the head occurs and $f_n(H)$ represents the relative frequency that the head occurs, $f_n(H)=\frac{f_H}{n}$.

Table 2.1

Experiment order	$n = 5$		$n = 50$		$n = 500$	
	f_H	$f_n(H)$	f_H	$f_n(H)$	f_H	$f_n(H)$
1	3	0.6	26	0.52	248	0.496
2	4	0.8	30	0.60	260	0.520
3	2	0.4	22	0.44	252	0.504
4	5	1	23	0.46	257	0.514
5	2	0.4	19	0.38	249	0.498
6	1	0.2	28	0.56	253	0.506
7	3	0.6	20	0.40	255	0.510
8	4	0.8	29	0.58	246	0.492
9	3	0.6	24	0.48	258	0.516
10	1	0.2	21	0.42	256	0.512

When the number of tosses n increases, the data we get are shown in Table 2.2.

Table 2.2

n	f_H	$f_n(H)$
1800	929	0.5161
5500	2787	0.5067
13000	6516	0.5012
25000	12505	0.5002

It can be seen from the above data:

(1) The relative frequency has the characteristics of random oscillation. In other words, for the same n, $f_n(H)$ we get is not necessarily the same;

(2) When the number of tosses n is small, the oscillation range of the relative frequency $f_n(H)$ is large. But as n increases, the relative frequency $f_n(H)$ becomes stable. That is, when n increases, $f_n(H)$ always oscillates around 0.5 and gradually stabilises at 0.5.

From this example, we can see that when n is relatively small, the relative frequency of the occurrence of event H oscillates randomly between 0 and 1. Therefore, it is inappropriate to use the relative frequency $f_n(H)$ to represent the possibility of the occurrence of event H. However, when n gradually increases, the relative frequency $f_n(H)$ gradually stabilises at a constant.

For any event A, there is always an objective constant corresponding to it. Human practice has continuously proven this characteristic of "relative frequency stability" (generally known as statistical law). It reveals the regularities hidden in random phenomena. It is appropriate for us to use this stable relative frequency value to represent an event's occurrence probability. That is,

$$\lim_{n\to\infty} f_n(A) = P(A) \qquad \text{When } n \text{ is large, } f_n(A) \approx P(A).$$

However, in practical applications, it is impossible for us to do multiple experiments for each event to obtain a stable value of relative frequency. At the same time, for the needs of theoretical research, inspired by the stable characteristics of relative frequency, we can give a definition of probability, which measures the possibility of occurrence of an event.

Definition of probability

Let S be a random experiment, and U be its sample space. A real number is given for each event A of S, denoted as P(A). It is called the probability of event A.

Probability (set function) P(•) satisfies the following three conditions:

(1) For each event A $0 \le P(A) \le 1$.

(2) $P(U) = 1$

(3) If $A_1, A_2, \ldots$ are mutually exclusive events, that is, for $i \neq j$, $A_i \cap A_j = \varnothing$, $i, j = 1, 2, \ldots$, then we have

$$P(A_1 \cup A_2 \cup \ldots) = P(A_1) + P(A_2) + \ldots = \sum_{i=1}^{\infty} P(A_i)$$

This equation embodies the characteristics of probability. That is, probabilities can be added and listed.

Classical probability (equally likely type)

If a random experiment S has the following characteristics:

1. the number of elements in the sample space of the experiment is finite.
2. the probability of occurrence of each simple event of the experiment is the same.

then it is an experiment of equally likely type. Many experiments have the above characteristics. It was the primary research object at the beginning of the theoretical development of probability theory, and therefore it is also called the classical probability type. Some concepts of equally likely type are simple to understand and widely used.

In the following, we discuss the calculation formula for events of equally likely type.

Let the sample space of experiment S be U = $\{e_1, e_2, ..., e_n\}$. Because the probability of occurrence of all simple events of experiment S is the same, that is,

$$P\{e_1\} = P\{e_2\} = ... = P\{e_n\}$$

Moreover, because all simple events are mutually exclusive ($\{e_i\} \cap \{e_j\} = \varnothing$)

so
$$\begin{aligned} 1 = P(U) &= P(\{e_1\} \cup \{e_2\} \cup ... \cup \{e_n\}) \\ &= P\{e_1\} + P\{e_2\} + ... + P\{e_n\} \\ &= nP\{e_i\} \qquad (n(U) = n) \end{aligned}$$

Hence
$$P\{e_i\} = \frac{1}{n} = \frac{1}{n(U)} \qquad (i = 1, 2, \ldots, n)$$

This equation shows the symmetry of sample space U --- all the simple events have the same probability.

If event A consists of K simple events, that is,

$$A = \{e_{i_1}\} \cup \{e_{i_2}\} \cup ... \cup \{e_{i_k}\} = \{e_{i_1}, e_{i_2}, ..., e_{i_k}\}$$

$$(1 \le i_1 \le i_2 \le ... \le i_k \le n)$$

then

$$P(A) = \sum_{j=1}^{k} P\{e_{i_j}\} = kP\{e_i\} = \frac{k}{n} = \frac{n(A)}{n(U)} = \frac{\textit{Number of simple events in A}}{\textit{Number of all simple events in U}}$$

where $n(A) = k, \quad n(U) = n.$

That is,
$$P(A) = \frac{n(A)}{n(U)} = \frac{k}{n}$$

This is the formula for calculating the probability of event A in an equally likely type experiment.

Example 1. Throw a die. Calculate the probability that the number obtained is even.

Solution. Let the event A = {the number obtained is even}.

$$U = \{1, 2, 3, 4, 5, 6\} \qquad n(U) = 6$$

$$A = \{2, 4, 6\} \qquad n(A) = 3$$

So $$P(A) = \frac{n(A)}{n(U)} = \frac{3}{6} = \frac{1}{2}$$

In this example, we list all elements of U and A. However, this is difficult and inconvenient when the number of elements of U is relatively large. Therefore, n(A) and n(U) are generally determined based on combination or the principle of combination multiplication.

Example 2. There are 4 white balls and 2 red balls in a box. The balls are drawn twice from the box, one ball at random each time. Consider two drawing methods.

(a) Draw a ball and observe its colour. Put it back in the box. Mix the balls well. Then, draw a ball. This method is called sampling with replacement.

(b) Draw a ball without putting it back in the box. Draw another ball from the remaining balls in the box. This method is called sampling without replacement.

Calculate the probability of the following outcomes for both methods.

(1) Both balls are white.

(2) Both balls are red.

(3) The first ball drawn is red. The ball drawn the second time is white.

Solution. **(a) Sampling with replacement**

Let A = event (1), B = event (2), C = event (3).

(1) Draw two balls from the box. Each draw is a simple event. For the first time, there are 6 balls to choose from. There are also 6 balls to choose from for the second time. According to the principle of combination multiplication, there are altogether $\binom{6}{1}\binom{6}{1}$ drawing ways. That is, $n(U) = \binom{6}{1}\binom{6}{1}$. For event A, there are 4 white balls available for drawing for the first time. There are still 4 white balls available for drawing for the second time. According to the principle of combination multiplication, we have $n(A) = \binom{4}{1}\binom{4}{1}$.

So
$$P(A) = \frac{n(A)}{n(U)} = \frac{\binom{4}{1}\binom{4}{1}}{\binom{6}{1}\binom{6}{1}} = \frac{4}{9}$$

(2) Using a principle similar to (1), we get
$$P(B) = \frac{n(B)}{n(U)} = \frac{\binom{2}{1}\binom{2}{1}}{\binom{6}{1}\binom{6}{1}} = \frac{1}{9}$$

(3)
$$P(C) = \frac{n(C)}{n(U)} = \frac{\binom{2}{1}\binom{4}{1}}{\binom{6}{1}\binom{6}{1}} = \frac{2}{9}$$

(b) Sampling without replacement

(1) $$P(A) = \frac{n(A)}{n(U)} = \frac{\binom{4}{1}\binom{3}{1}}{\binom{6}{1}\binom{5}{1}} = \frac{2}{5}$$

(2) $$P(B) = \frac{n(B)}{n(U)} = \frac{\binom{2}{1}\binom{1}{1}}{\binom{6}{1}\binom{5}{1}} = \frac{1}{15}$$

(3) $$P(C) = \frac{n(C)}{n(U)} = \frac{\binom{2}{1}\binom{4}{1}}{\binom{6}{1}\binom{5}{1}} = \frac{4}{15}$$

Example 3. There are 10 balls in a bag, including 4 white balls and 6 red balls. 5 balls are drawn simultaneously. What is the probability of getting 2 white balls and 3 red balls?

Solution. Let the event A = {Get 2 white balls and 3 red balls}.
According to the principle of combination multiplication

$$n(U) = \binom{10}{5} \qquad n(A) = \binom{4}{2}\binom{6}{3}$$

so $$P(A) = \frac{n(A)}{n(U)} = \frac{\binom{4}{2}\binom{6}{3}}{\binom{10}{5}} = \frac{10}{21}$$

Example 4. The box contains 3 black balls, 4 white balls, and 5 red balls. Take at random 5 balls. What is the probability of getting 2 red balls and 2 black balls?

Solution. Let the event A = {Get 2 red balls and 2 black balls}.

According to the principle of combination multiplication

$$n(\mathrm{U}) = \binom{12}{5} \qquad\qquad n(\mathrm{A}) = \binom{5}{2}\binom{3}{2}\binom{4}{1}$$

so
$$\mathrm{P(A)} = \frac{n(\mathrm{A})}{n(\mathrm{U})} = \frac{\binom{5}{2}\binom{3}{2}\binom{4}{1}}{\binom{12}{5}} = \frac{5}{33}$$

Example 5. Throw n balls randomly into N ($\mathrm{N} \geq n$) boxes. Find the probability that each box contains at most one ball (the capacity of the box is not limited).

Solution. Throw n balls into N boxes randomly. Each ball can be thrown into any of the N boxes. Throw n times. So

$$n(\mathrm{U}) = \binom{\mathrm{N}}{1}\binom{\mathrm{N}}{1}\cdots\binom{\mathrm{N}}{1} = \mathrm{N}^n$$

Let A = {each box contains at most one ball}, then

$$n(\mathrm{A}) = \binom{\mathrm{N}}{1}\binom{\mathrm{N}-1}{1}\cdots\binom{\mathrm{N}-(n-1)}{1} = \mathrm{N}(\mathrm{N}-1)\cdots[\mathrm{N}-(n-1)] = {}^{\mathrm{N}}P_n$$

where ${}^{\mathrm{N}}P_n$ represents the number of permutations.

Therefore, the probability that each box contains at most one ball is

$$\mathrm{P(A)} = \frac{n(\mathrm{A})}{n(\mathrm{U})} = \frac{\binom{\mathrm{N}}{1}\binom{\mathrm{N}-1}{1}\cdots\binom{\mathrm{N}-(n-1)}{1}}{\binom{\mathrm{N}}{1}\binom{\mathrm{N}}{1}\cdots\binom{\mathrm{N}}{1}} = \frac{{}^{\mathrm{N}}P_n}{\mathrm{N}^n}$$ □

Many problems have the same mathematical model as this example. For example, assuming everyone's birthday is equally possible on any day of 365 days in a year, then randomly select n people ($n \leq 365$). The probability that their birthdays are different is

$$\frac{{}^{N}P_n}{N^n} = \frac{365 \cdot 364 \cdots (365-n+1)}{365^n}$$

Therefore, the probability that at least two people among n people have the same birthday is

$$P = 1 - \frac{365 \cdot 364 \cdots (365-n+1)}{365^n}$$

After calculation, the following results can be obtained.

n	20	30	40	50	64	100
P	0.411	0.706	0.891	0.970	0.997	0.9999997

The table shows that when $n = 64$, the probability of "at least two people having the same birthday" is $P \to 1$, so it can be considered a definite event.

Example 6. A consulting agency had 15 visitors in one week. All 15 visits are known to have occurred on two days, Wednesday and Friday. Is it possible to infer that the consulting agency has specified reception hours?

Solution. Suppose there are no specified reception hours at the consulting agency. In that case, each visitor has an equal probability of going to the consulting agency on any day of the seven days a week. The consulting agency had 15 visitors a week, which is equivalent to 15 draws, then

$$n(U) = \binom{7}{1}\binom{7}{1}\cdots\binom{7}{1} = 7^{15}$$

Let A = {All 15 visits were on Wednesday and Friday}, then

$$n(A) = \binom{2}{1}\binom{2}{1}\cdots\binom{2}{1} = 2^{15}$$

Therefore, if there are no specified reception hours at the consulting agency, the probability that all 15 receptions are on Wednesday and Friday is

$$P(A)=\frac{n(A)}{n(U)}=\frac{2^{15}}{7^{15}}=6.90\times10^{-9}$$

People have concluded from long-term practice that events with a very small probability ($P\to0$) will not occur in a trial, so it can be inferred that the consulting agency does not receive visitors daily. That is, the reception time is specified.

Example 7. Toss a coin once. Find the probability of getting a head and getting a tail.

Solution. Let H = head, T = tail.

Judging from the actual situation, the probability of getting a head or a tail is equal in a coin toss experiment. That is,

$$P\{H\}=P\{T\}$$

Since $\quad U=\{H,\ T\},\ \ n(U)=\binom{2}{1},\quad n(H)=\binom{1}{1},\quad n(T)=\binom{1}{1}$

so $\quad P\{H\}=\dfrac{n(H)}{n(U)}=\dfrac{\binom{1}{1}}{\binom{2}{1}}=\dfrac{1}{2}\qquad P\{T\}=\dfrac{n(T)}{n(U)}=\dfrac{\binom{1}{1}}{\binom{2}{1}}=\dfrac{1}{2}$

That is, $\quad P\{H\}=P\{T\}=\dfrac{1}{2}$

What needs to be pointed out is that when a coin is tossed once, we have enough known information to calculate the probability directly. For the relative frequency mentioned earlier, when there is insufficient known information, it can be used to approximate the probability P(A).

For example, pick up an apple randomly from a pile of apples. What is the probability of getting a good apple?

Solution. Since the total number of all apples and the number of good and bad apples are unknown, that is, there is no information we need. Therefore, we can only use the relative frequency method to obtain an approximate value of probability P(A).

Because $$\mathrm{P(A)} = \lim_{n\to\infty} f_n(\mathrm{A})$$

when n is relatively big,

$$\mathrm{P(A)} \approx f_n(\mathrm{A})$$

We randomly draw $n = 20$ apples, and we get $f_\mathrm{A} = 14$ good apples.

So $$\mathrm{P(A)} \approx f_n(\mathrm{A}) = \frac{f_A}{n} = \frac{14}{20} = 0.7$$

That is, $$\mathrm{P(A)} \approx 0.7$$

The probability value also tells us the proportion of good and bad apples. In other words, in this pile of apples, 70% are good apples, and 30% are bad apples.

In summary:

1. n(U) is determined according to the given condition of U and the experiment method. n(A) is determined according to the given condition of A and the experiment method.

2. We can use a simple example to test the correctness of determining n(U) and n(A) with combination or the principle of combination multiplication. For example, there are 3 balls in a bag. 2 are red and 1 is white. The balls are drawn twice, one ball at random each time. The drawing is sampling without replacement. What is the probability of getting two red balls?

   Solution. Number the three balls. Number the two red balls (1), (2) and the white ball 3. (1)3 means that the first ball obtained is the red ball No. (1), and the second ball is white ball No. 3. According to the sampling method, here are all simple events for U and A.

$$U = \{(1)(2), (1)3, (2)(1), (2)3, \ 3(1), 3(2)\} \qquad n(U) = 6 = \binom{3}{1}\binom{2}{1}$$

$$A = \{(1)(2), (2)(1)\} \qquad n(A) = 2 = \binom{2}{1}\binom{1}{1}$$

So $$P(A) = \frac{n(A)}{n(U)} = \frac{\binom{2}{1}\binom{1}{1}}{\binom{3}{1}\binom{2}{1}} = \frac{2}{6} = \frac{1}{3}$$

This example illustrates that it is correct to use combination or the principle of combination multiplication to determine $n(U)$ and $n(A)$ in $P(A) = \frac{n(A)}{n(U)}$. It avoids the trouble of listing all elements of U and A.

Exercise 2.1

1. A bag has 20 cards numbered from 1 to 20. A card is drawn at random. List the sample space and find the probability that the number on the card is

a. an odd number **b.** a prime number **c.** divisible by 4 or 5

2. A letter is chosen at random from the word DISTRIBUTION. Find the probability that the letter chosen is

a. an I **b.** a consonant **c.** a vowel

3. A box contains 3 marbles: 1 black, 1 red, and 1 pink. Take 1 marble randomly from the box, observe its colour, and return it to the box. Then, randomly draw another marble from the box.

a. List the sample space.

b. Find the probability that the colours of the first and the second marbles are different.

4. A coin is tossed three times. List the sample space and find the probability of

a. 3 heads **b.** 2 heads and 1 tail **c.** 1 head

d. no head **e.** at least one head **f.** at most 1 head

5. A box contains 4 cards numbered 5, 6, 7, 8. Two cards are drawn at random without replacement one after another from the box. Let A = "The largest number drawn is 7".

a. List the sample space. **b.** Find P(A).

6. A class consists of 10 boys and 6 girls. A committee of 3 is chosen at random from the class. Find the probability that

a. 3 boys are selected; **b.** 2 boys are selected;

c. at least one boy is selected; **d.** 2 girls are selected.

7. A family has three children. List the sample space for the sexes of the children from the oldest to the youngest. Find the probability of the following events.

a. two boys; **b.** more boys than girls; **c.** the oldest is a boy;

d. no girl younger than a boy; **e.** at most one boy.

8. An urn contains 7 yellow balls and 5 black balls. A ball is chosen randomly from the urn and returned together with 3 more balls of the same colour. Then, a second ball is drawn at random from the urn. Find the probability that

a. both balls are yellow;

b. the first ball is yellow, and the second ball is black;

c. both balls are of the same colour.

9. A box contains 10 white balls and 5 golden balls. Take 3 balls at random from the

box. Find the probability that the 3 balls are of the same colour.

10. A deck of a standard 52 playing cards consists of 13 red diamonds (♦), 13 red hearts (♥), 13 black clubs (♣) and 13 black spades (♠). Two cards are drawn at random, one after the other from the deck. Calculate the probability that

a. both cards are hearts;

b. the first card is red, and the second card is black;

c. both cards are of the same colour.

11. There are 2 defective transistors out of 10. Take two of them randomly, one at a time, using sampling without replacement. Find the probability of the following events

a. both are non-defective;

b. both are defective;

c. one is non-defective and one is defective;

d. the second one is defective.

12. Take at random a phone number from a phone number book. Each of the last four digits has an equal probability of having a value between 0 and 9. Find the probability that the last four digits are all different.

13. There are 10 people in the room, wearing medals numbered 1 to 10. Choose at random 3 people to record the numbers of their medals. Find the probability that

a. the smallest number is 5;

b. the biggest number is 5.

14. Randomly select 3 numbers from 0 to 9 without replacement and line them up to form a 3-digit number. Find the probability of forming a 3-digit even number.

15. **a.** Randomly arrange the letters a, b, b, i, i, l, o, p, r, t, y. Find the probability that it exactly spells the word "probability".

b. Choose at random 7 letters from the 11 letters in **a**. and arrange them. Find the probability that the result of the arrangement is the word "ability".

2.2. Relationship between events and calculation of events

An event is a set. Therefore, the relationship between events and the calculation of events are naturally handled in accordance with the relationship between sets and the operation of sets in the set theory. At the same time, according to the definition of "occurrence of event", we will give their meaning in probability theory.

1. Probability of complementary events

If $A \cup A' = U$ and $A \cap A' = \varnothing$, then we say that event A and event A' are complementary events. That is, the complementary event of A is A'. We can express it as A' or $\bar{A}$.

$$1 = P(U) = P(A \cup A') = P(A) + P(A')$$

That is, $P(A) + P(A') = 1$ (**complementary rule**)

Or $P(A) + P(\bar{A}) = 1$

Venn diagram

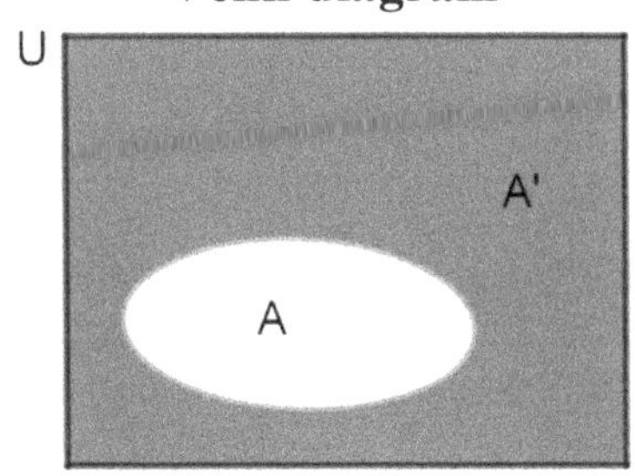

Graph 2.1

2. Probability of the sum event (union event)

Event $C = A \cup B = \{x \mid x \in A \text{ or } x \in B\}$, and it is called the **sum event** (or **union event**) of event A and event B. $C = A \cup B$ occurs if and only if at least one of events A and B occur.

From the operation of sets

$$A \cup B = (A \cap B) \cup (A \cap B') \cup (B \cap A') \qquad \text{we get}$$

$$\text{The sum event C occurs} = (A \cup B) \text{ occurs} = \begin{cases} \textit{Event} \quad (A \cap B) \quad \textit{occurs.} \\ \qquad\qquad \textit{or} \\ \textit{Event} \quad (A \cap B') \quad \textit{occurs.} \\ \qquad\qquad \textit{or} \\ \textit{Event} \quad (B \cap A') \quad \textit{occurs.} \end{cases}$$

By adding the probabilities of the three possibilities, we get the **probability of the sum event C.**

$$P(C) = P(A \cup B) = P(A \cap B) + P(A \cap B') + P(B \cap A')$$
$$= P(A) + P(B) - P(A \cap B) \qquad \textbf{(addition rule)}$$

In practice, sometimes it is more convenient to understand as follows.

$$\text{The sum event C occurs} = (A \cup B) \text{ occurs} = \begin{cases} \textit{Event} \quad A \quad \textit{occurs.} \\ \qquad\quad \textit{or} \\ \textit{Event} \quad B \quad \textit{occurs.} \\ \qquad\quad \textit{or} \\ \textit{Event} \quad A \cap B \quad \textit{occurs.} \end{cases}$$

where event $(A \cap B)$ occurs = event A and event B occur at the same time.

So $\qquad P(C) = P(A) + P(B) - P(A \cap B)$

This can be seen as the addition of three probabilities. But it should be noted that the symbol before $P(A \cap B)$ is '−'.

Venn diagram of the sum event $C = A \cup B$

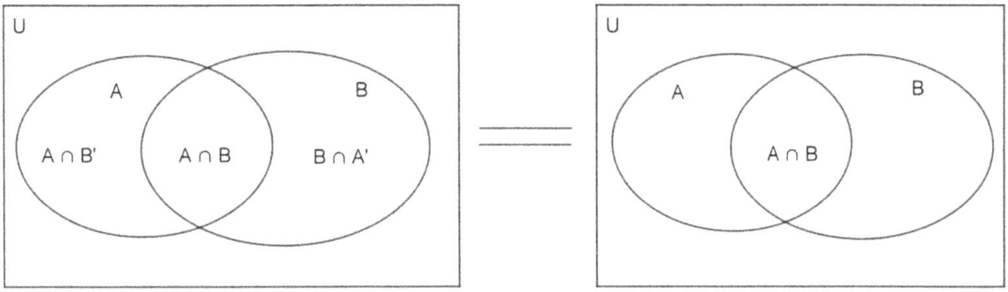

Graph 2.2

Operation of sets

$$A \cup B = (A \cap B') \cup (A \cap B) \cup (B \cap A') = A \cup (B - A \cap B)$$

To facilitate the understanding of the meaning of the sum event, we give an easy-to-understand example.

$$\text{We getting rich (C) occurs} = \begin{cases} \textit{winning the first prize of lottory } (\text{A}) \textit{ occurs} \\ \textit{or} \\ \text{becoming } \textit{the CEO of Google } (\text{B}) \textit{ occurs} \\ \textit{or} \\ \text{A } \textit{and } \text{B } \textit{occur } \text{simultaneously} \end{cases}$$

So the possibility of getting rich, that is, the probability of the sum event C, is

$$P(C) = P(A) + P(B) - P(A \cap B)$$

If A and B do not occur at the same time, that is,

$$P(A \cap B) = P(\varnothing) = 0$$

then the possibility of getting rich, that is, the probability of the sum event C, is

$$P(C) = P(A) + P(B)$$

3. Probability of the product event (intersection event)

Event $D = A \cap B = \{x \mid x \in A \text{ and } x \in B\}$, and it is called the **product event** (or **intersection event**) of event A and event B. If and only if event A and event B both occur (including the case where one event occurs first and the other occurs later), event $D = A \cap B$ occurs. We can express this as $A \cap B$, or A, B or AB.

The product event D occurs = $(A \cap B)$ occurs = Event A and event B both occur.

Probability of the product event $D = A \cap B$ (multiplication rule)

Let $P(A) > 0$, then

$$P(D) = P(A \cap B) = P(A)P(B \mid A) = P(B)P(A \mid B)$$

where $P(B \mid A)$, $P(A \mid B)$ are conditional probabilities.

Venn diagram of the product event $D = A \cap B$

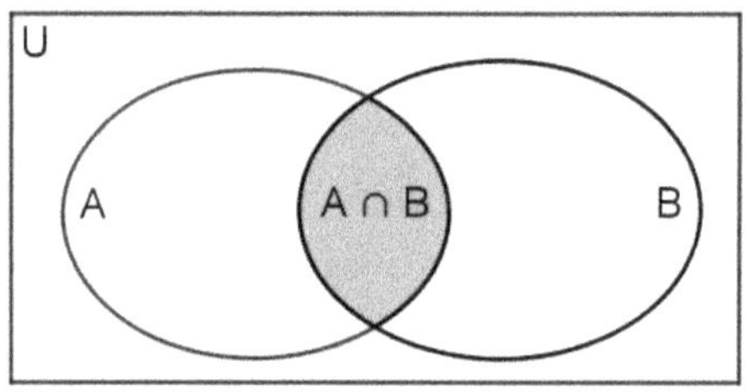

Graph 2.3

If $D = A \cap B = \varnothing$, then we say that event A and event B are mutually exclusive. That is, event A and event B cannot occur at the same time.

$$P(D) = P(A \cap B) = P(\varnothing) = 0$$

Venn diagram

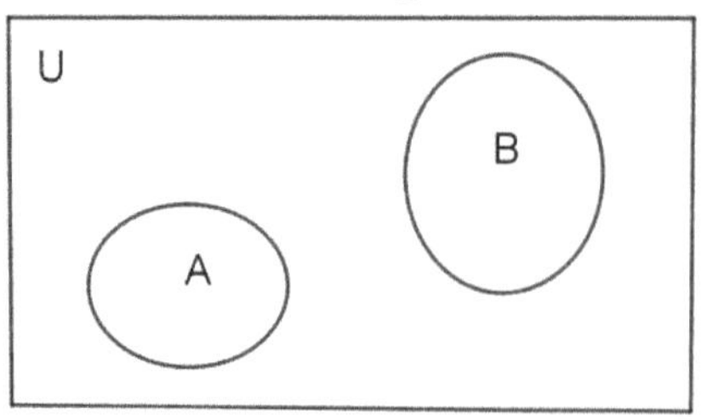

$A \cap B = \varnothing$

Graph 2.4

If $A_1, A_2, \ldots, A_n$ are mutually exclusive events, that is, every two events cannot occur at the same time – $A_i \cap A_j = \varnothing$ ($i \neq j$, $i, j = 1, 2, \ldots, n$), for the sum event C_n of $A_1, A_2, \ldots, A_n$, $C_n = A_1 \cup A_2 \cup \ldots \cup A_n$, we have

$$C_n \text{ occurs} = (A_1 \cup A_2 \cup \ldots \cup A_n) \text{ occurs} = \begin{cases} A_1 & \textit{occurs.} \\ \textit{or} & \\ A_2 & \textit{occurs.} \\ \textit{or} & \\ \vdots & \\ \textit{or} & \\ A_n & \textit{occurs.} \end{cases}$$

By adding n probabilities, we can get the probability of the sum event C_n.

$$P(C_n) = P(A_1 \cup A_2 \cup \ldots \cup A_n) = P(A_1) + P(A_2) + \ldots + P(A_n)$$

This equation shows that probability can be listed and added.

Exercise 2.2

1. A die is tossed 3 times. Find the probability that

a. a 5 on the first toss **b.** a 5 on at least one of these tosses.

2. A small school has 200 students. 100 of them study French. 60 study German and 40 study both French and German. A student is chosen at random. Find the probability that the student

a. studies none of these two languages;

b. studies at least one of these two languages;

c. studies German but not French.

3. The events A and B are such that $P(A) = 0.6$, $P(B) = 0.8$, $P(A \cap B) = 0.5$. Find

a. $P(A \cup B)$ **b.** $P(A')$ **c.** $P(B' \cap A)$

4. a. $P(A \cap B) = 0.2$, $P(A \cup B) = 0.7$, $P(A) = 0.4$. Find $P(B)$.

b. $P(A) = 0.5$, $P(B) = 0.2$. A and B are mutually exclusive. Find $P(A \cup B)$.

c. $P(A) = 0.5$, $P(B) = 0.5$, $P(\overline{A \cap B}) = 0.6$. Find $P(B \cap A')$.

5. The events A and B are such that $P(A) = 0.55$, $P(B) = 0.7$, $P(A \cap B) = 0.45$. Find (using a Venn diagram if appropriate)

a. $P(A \cup B)$ **b.** $P(\overline{A \cap B})$ **c.** $P(A \cup B')$ **d.** $P(A' \cap B')$

6. Let A and B be events with $P(A \cup B) = \frac{3}{4}$, $P(A') = \frac{2}{3}$, $P(A \cap B) = \frac{1}{4}$. Find

a. $P(A)$ **b.** $P(B)$ **c.** $P(A \cap B')$

7. The events A and B are such that $P(A) = 0.7$, $P(B) = 0.8$.

a. Under what condition will $P(A \cap B)$ have a maximum value? What is the maximum value?

b. Under what condition will $P(A \cap B)$ have a minimum value? What is the minimum value?

8. 2 items are chosen randomly from 100 items, of which 3 are defective. Find the probability that

a. none is defective; **b.** both are defective; **c.** at least one is defective.

9. A class has 10 boys and 20 girls. Half the boys and half the girls have blue eyes. Find the probability that a person chosen at random is a girl or has blue eyes.

10. In a certain community, 60% of households own a car, and 30% own a house.

20% of households own both a car and a house. A household is selected at random. Find the probability that this family owns a car or a house but not both.

11. 40% of the residents of a small town subscribe to newspaper A, 25% of the residents subscribe to newspaper B. 15% of the residents subscribe to both newspapers. A resident is chosen at random. Find the probability that the resident

a. subscribes to only newspaper A;

b. subscribes to only one newspaper;

c. subscribes to at least one newspaper;

d. do not subscribe to any newspaper.

12. There are two bags, A and B. Bag A contains n white balls and m red balls. Bag B contains N white balls and M red balls. Take a ball from bag A at random and put it into bag B, and then take a ball from bag B at random. Find the probability of getting a white ball.

2.3. Conditional probability and Independence

Conditional probability

In the previous section, we have studied the probability of the product event $P(A \cap B) = P(A)P(B \mid A)$, where $P(B \mid A)$ is the conditional probability. Conditional probability is an important concept in probability theory. It studies the probability of occurrence of event B, given that event A has already occurred.

Let us read one example. Toss a coin twice and observe the occurrence of head and tail. Let the event A = {Get at least one head}, event B = {Both tosses have the same result}. Calculate the probability of event B, provided event A has already occurred.

Here, the sample space U = {HH, HT, TH, TT}, where H = head, T = tail.

$$n(\text{U}) = 4$$

$\text{A} = \{\text{HH, HT, TH}\}$, $n(\text{A}) = 3$, $\text{B} = \{\text{HH, TT}\}$, $n(\text{B}) = 2$, $n(\text{A} \cap \text{B}) = 1$

It is given that event A has already occurred. Therefore, given that event A has already occurred, the probability of event B is

$$\text{P}(\text{B} \mid \text{A}) = \frac{n(\text{A} \cap \text{B})}{n(\text{A})} = \frac{1}{3}$$

Here we can find $\quad \text{P}(\text{B}) = \dfrac{n(\text{B})}{n(\text{U})} = \dfrac{2}{4} \neq \text{P}(\text{B} \mid \text{A})$

This is easy to understand because when we calculate P(B | A), we have set the condition to consider the occurrence probability of event B when event A has already occurred.

On the other hand, it is easy to get

$$\text{P}(\text{A}) = \frac{n(\text{A})}{n(\text{U})} = \frac{3}{4} \qquad \text{P}(\text{A} \cap \text{B}) = \frac{n(\text{A} \cap \text{B})}{n(\text{U})} = \frac{1}{4}$$

So $\quad \text{P}(\text{B} \mid \text{A}) = \dfrac{1}{3} = \dfrac{\frac{1}{4}}{\frac{3}{4}} = \dfrac{\text{P}(\text{A} \cap \text{B})}{\text{P}(\text{A})} = \dfrac{n(\text{A} \cap \text{B})}{n(\text{A})}$

That is, $\quad \text{P}(\text{B} \mid \text{A}) = \dfrac{\text{P}(\text{A} \cap \text{B})}{\text{P}(\text{A})} = \dfrac{n(\text{A} \cap \text{B})}{n(\text{A})}$

Definition of conditional probability

Let A and B be two events. P(A) > 0.

We call $\quad \text{P}(\text{B} \mid \text{A}) = \dfrac{\text{P}(\text{A} \cap \text{B})}{\text{P}(\text{A})}$

as the conditional probability of occurrence of event B given that event A has occurred.

Note:

P(B|A) is the probability of event $B \mid A$. In addition to using the definition equation, we can also use the formula $P(B|A) = \dfrac{n(B|A)}{n(U)}$ to find it. This depends on the situation.

It is easy to prove that the conditional probability $P(\bullet \mid A)$ satisfies the three conditions in the definition of probability. That is,

1. For each event B, we have $0 \le P(B \mid A) \le 1$.

2. $P(U \mid A) = 1$

   That is, $1 = P(U \mid A) = P(B \cup B' \mid A) = P(B \mid A) + P(B' \mid A)$

   So $1 = P(B \mid A) + P(B' \mid A)$

3. Let $B_1, B_2, \ldots$ be mutually exclusive events $(B_i \cap B_j = \varnothing, i \neq j)$. Then we have

$$P(B_1 \mid A \cup B_2 \mid A \cup \ldots) = \sum_{i=1}^{\infty} P(B_i|A)$$

Example. There are 5 products in a box, 3 of which are first-class and 2 second-class. Take products from the box twice, one piece at a time, without replacement.
Let the event A = {obtaining a first-class product for the first time},
the event B = {obtaining a first-class product for the second time}.
Find the conditional probability P(B|A).

Solution. Method 1 Sampling without replacement

The probability of getting the first-class products twice, that is, the probability that A and B both occur, is

$$P(A \cap B) = P(A)P(B|A) = \frac{\binom{3}{1}}{\binom{5}{1}} \cdot \frac{\binom{2}{1}}{\binom{4}{1}} = \frac{3}{10}$$

So
$$P(B|A) = \frac{P(A \cap B)}{P(A)} = \frac{\frac{3}{10}}{\binom{3}{1}\Big/\binom{5}{1}} = \frac{1}{2}$$

Method 2 Sampling without replacement

Under the condition that the first-class product is obtained for the first time, the conditional probability of obtaining the first-class product for the second time is

$$P(B|A) = \frac{n(B|A)}{n(U)} = \frac{\binom{3-1}{1}}{\binom{5-1}{1}} = \frac{2}{4} = \frac{1}{2}$$

Independence

Let A and B be the two events of experiment S. If $P(A) > 0$, we can define that

$$P(B \mid A) = \frac{P(A \cap B)}{P(A)}$$

Generally speaking, if the occurrence of A will affect the probability of the occurrence of B, then

$$P(B \mid A) \neq P(B)$$

and we say that A and B are dependent. Only when this influence does not exist, we have

$$P(B \mid A) = P(B)$$

At this time, we have

$$P(A \cap B) = P(A)P(B \mid A) = P(A)P(B)$$

And we say that A and B are independent.

For example, two coins are numbered coin No. 1 and coin No. 2, respectively. Let experiment S be "tossing coins No. 1 and No. 2 simultaneously and observing the heads and tails obtained." Let the event A = {Coin No. 1 gets a head}, B = {Coin No. 2 gets a head}.

The sample space U of the experiment S = {HH, HT, TH, TT}, $n(U) = 4$

$$A = \{HH, HT\} \qquad n(A) = 2$$

$$B = \{HH, TH\} \qquad n(B) = 2 \qquad n(A \cap B) = 1$$

So $$P(A) = \frac{n(A)}{n(U)} = \frac{2}{4} = \frac{1}{2} \qquad P(B) = \frac{n(B)}{n(U)} = \frac{2}{4} = \frac{1}{2}$$

$$P(B \mid A) = \frac{P(A \cap B)}{P(A)} = \frac{n(A \cap B)}{n(A)} = \frac{1}{2} \qquad P(A \cap B) = \frac{n(A \cap B)}{n(U)} = \frac{1}{4}$$

We can find out here that

$$P(B \mid A) = P(B) \qquad P(A \cap B) = P(A)P(B)$$

It is evident from the known information that whether coin No. 1 comes up heads does not affect whether coin No. 2 comes up heads.

Definition

Let A and B be two events. If we have

$$P(A \cap B) = P(A)P(B)$$

then A and B are independent events.

It is easy to prove

1. If events A and B are independent, then A, A', and B' are mutually independent. B, A' and B' are mutually independent of each other;
2. If event A and event B are independent, $A \cap B \neq \varnothing$, $P(A) > 0$, $P(B) > 0$, then

$$P(A \cap B) = P(A)P(B) > 0$$

Theorem

Let A and B be two events. If event A and event B are independent, $P(A) > 0$, $P(B) > 0$, then

$$P(B \mid A) = P(B), \qquad P(A \mid B) = P(A)$$

In practice, we generally do not judge the independence of events based on the definition. However, we will determine this based on the practical implications.

Below, we prove that if event A and event B are mutually independent, then

(a) A' and B are mutually independent.

(b) A and B' are mutually independent.

(c) A' and B' are mutually independent.

Proof: (a) Since A and B are mutually independent, hence

$$P(A \mid B) = P(A)$$

And $$P(A') = 1 - P(A) = 1 - P(A \mid B) = P(A' \mid B)$$

That is, $$P(A') = P(A' \mid B)$$

Therefore, A' and B are mutually independent.

(b) Since A and B are mutually independent, hence

$$P(B \mid A) = P(B)$$

And

$$P(A \cap B') = P(B' \mid A)P(A) = [1 - P(B \mid A)]P(A) = [1 - P(B)]P(A) = P(B')P(A)$$

Therefore, A and B' are mutually independent.

(c) Method 1

$$P(A' \cap B') = P(A' \mid B')P(B') = [1 - P(A \mid B')]P(B')$$

Using the result of (b), A and B' are mutually independent, we get

$$P(A' \cap B') = [1 - P(A \mid B')]P(B') = [1 - P(A)]P(B') = P(A')P(B')$$

Hence, A' and B' are mutually independent.

Method 2

Because $\quad P(A' \mid B') + P(A \mid B') = 1$

so $\quad P(A' \mid B') = 1 - P(A \mid B') = 1 - P(A) = P(A')$

Therefore, A' and B' are mutually independent.

*Finally, three extended concepts need to be added.

1. Probability of the sum event of *n* events $C = \bigcup_{k=1}^{n} A_k$

If A_1, A_2, and A_3 are three events, then we have

$$P(A_1 \cup A_2 \cup A_3) = P(A_1) + P(A_2) + P(A_3) - P(A_1A_2) - P(A_1A_3) - P(A_2A_3) + P(A_1A_2A_3)$$

If A_1, A_2, A_3 are mutually exclusive events, that is, $i \neq j$, $A_i \cap A_j = \varnothing$, $i, j = 1, 2, 3$, then we have

$$P(A_1 \cup A_2 \cup A_3) = P(A_1) + P(A_2) + P(A_3) = \sum_{i=1}^{3} P(A_i)$$

where $C = \bigcup_{k=1}^{3} A_k = A_1 \cup A_2 \cup A_3$.

Generally, for n events $A_1, A_2,\ldots, A_n$, we can use induction to get

$$P(A_1 \cup A_2 \cup \ldots \cup A_n)$$

$$= \sum_{i=1}^{n} P(A_i) - \sum_{1 \le i<j \le n} P(A_i A_j) + \sum_{1 \le i<j<k \le n} P(A_i A_j A_k) + \ldots + (-1)^{n-1} P(A_1 A_2 \ldots A_n)$$

If $A_1, A_2,\ldots, A_n$ are mutually exclusive events, that is, $i \neq j,\ A_i \cap A_j = \varnothing$, $i, j = 1, 2,\ldots, n$, then we have

$$P(A_1 \cup A_2 \cup \ldots \cup A_n) = P(A_1) + P(A_2) + \ldots + P(A_n) = \sum_{i=1}^{n} P(A_i)$$

where $C = \bigcup_{k=1}^{n} A_k = A_1 \cup A_2 \cup \ldots \cup A_n$.

2. Probability of the product event of n events $D = \bigcap_{k=1}^{n} A_k$

If A_1, A_2, A_3 are three events, and $P(A_1A_2) > 0$, then we have

$$P(A_1A_2A_3) = P(A_3 \mid A_1A_2)P(A_2 \mid A_1)P(A_1)$$

When the events A_1, A_2, and A_3 are mutually independent, the equation changes to

$$P(A_1A_2A_3) = P(A_1)\, P(A_2)\, P(A_3)$$

where $D = \bigcap_{k=1}^{3} A_k = A_1A_2A_3$.

Generally, for n events $A_1, A_2,\ldots, A_n$ $(n \geq 2)$, and $P(A_1A_2\ldots A_{n-1}) > 0$, we have

$$P(A_1A_2\ldots A_n) = P(A_n \mid A_1A_2\ldots A_{n-1})P(A_{n-1} \mid A_1A_2\ldots A_{n-2}) \cdots P(A_2 \mid A_1)P(A_1)$$

When the events $A_1, A_2,\ldots, A_n$ are mutually independent, the equation changes to

$$P(A_1A_2\ldots A_n) = P(A_1)P(A_2) \cdots P(A_n)$$

where $D = \bigcap_{k=1}^{n} A_k = A_1A_2 \ldots A_n$.

3. Conditional independence of events

If under the condition that B occurs, the probability of A_1 occurring will not change regardless of whether A_2 occurs or not, then we say that A_1 and A_2 are conditionally independent given B.

That is, $P(A_1|\ A_2B) = P(A_1|\ B)$

or equivalently

$$P(A_1A_2|B) = P(A_1|\ B)P(A_2|\ B)$$

The concept of conditional independence can easily be extended to the case of more than two events, so it is omitted here.

Exercise 2.3

1. Toss two fair coins. Given that at least 1 head is shown, what is the probability that 2 heads are shown?

2. A green die and a red die are each tossed once.

- **a.** Given that the red die shows a 6, find the probability of obtaining a sum greater than 9.
- **b.** Given that the green die shows a 2, find the probability that the sum is less than 6.
- **c.** Given that the red die shows a number less than 5, find the probability that the sum is 11.
- **d.** Given that the sum is 7, find the probability that one of the dice shows 1.

3. A coin is tossed until a head appears or has been tossed three times. Given that the head does not occur on the first toss, what is the probability that the coin is tossed three times?

4. A class has 10 boys and 5 girls. Three students are randomly selected from the class, one after the other. Find the probability of the following

a. the first two are boys, and the third is a girl.

b. the first and the third are boys, and the second is a girl.

c. the first and third are of the same sex, and the second is of the opposite sex.

d. If the first and third are of the same sex and the second is of the opposite sex, what is the probability that the second is a girl?

5. Urn A contains 3 red balls and 6 white balls. Urn B contains 4 red balls and 4 white balls. Choose one of the urns at random and draw a ball.

a. What is the probability that it is a red ball?

b. Given that it is a red ball, what is the probability it is from urn A?

6. Given that $P(A) = \frac{1}{4}$, $P(B|A) = \frac{1}{3}$, $P(A|B) = \frac{1}{2}$, find $P(A \cup B)$.

7. The probability that at least one of event A and event B occurs is 0.12. The probability that event A and event B occur simultaneously is 0.1. Are event A and event B independent?

8. Someone asks a question. The rule is that A answers first. The probability that A gets it right is 0.4. If the answer is correct, the quiz is over. If the answer is wrong, B answers next, and after that, the quiz is over. It is known that B's answer and A's answer are independent. The probability that both of them get it wrong is 0.2. Find the probability that the question is answered correctly by B.

9. For one mathematics problem, in half an hour, the probability that Alice can solve it is $\frac{1}{2}$ and the probability that Betty can solve it is $\frac{1}{3}$. Alice and Betty try to solve the problem independently in half an hour. Find the probability that

a. none of them can solve the problem.

b. the problem is solved by Alice only.

c. the problem is solved.

d. Given that the problem is solved, what is the probability that Alice solves it?

10. Martin has to pass through 4 traffic lights from home to school. The probability of a red light appearing at traffic light ports 1 and 3 is 0.4. The probability of a red light appearing at traffic light ports 2 and 4 is 0.6. All traffic lights are independent of each other. Find the probability that he encounters at least 2 red lights from home to school.

11. Toss a fair coin three times. Let A = {at most one head shown}. B = {There are head(s) and tail(s) in three tosses}. Prove that events A and B are independent.

12. Two boxes contain items of the same type. The first box contains 50 items, of which 10 are first-class. The second box contains 30 items, of which 18 are first-class. Choose one box randomly and take two items randomly from the box, one after the other. Find the probability that

a. the first item taken is first-class.

b. Given that the first item taken is first-class, find the probability that the second item is also first-class.

13. Let A and B be events with $P(A) = \frac{1}{4}$, $P(A \cup B) = \frac{1}{3}$. $P(B) = p$. Find the value of p

a. if A and B are mutually exclusive.

b. if A and B are independent.

c. if A is a subset of B.

14. Let A and B be events with $P(A) = \frac{1}{2}$, $P(B) = \frac{1}{3}$, $P(A \cap B) = \frac{1}{4}$. Find

a. $P(A|B)$ **b.** $P(B|A)$ **c.** $P(A \cup B)$ **d.** $P(A'|B')$

e. $P(A|B')$

15. The probability that a man will live over 80 years old is $\frac{1}{4}$ and the probability that his wife will live over 80 years old is $\frac{1}{3}$. Find the probability that

a. at least one will live over 80 years old.

b. exactly one will live over 80 years old.

c. If exactly one will live over 80 years old, what is the probability that it will be the wife?

16. A and B independently decipher one password. The probabilities that they can decipher the password are $\frac{1}{3}$ and $\frac{1}{4}$, respectively. Find the probability that

a. both decipher the password.

b. none of them deciphers the password.

c. exactly one of them deciphers the password.

d. at most one of them deciphers the password.

e. If the probability of deciphering the password is $\frac{99}{100}$, how many people like B are needed at least?

17*. Let A, B, and C be three events such that $P(A) = P(B) = P(C) = \frac{1}{4}$,

$P(A \cap B) = P(B \cap C) = 0$, and $P(A \cap C) = \frac{1}{8}$. Find the probability that at least one of the events A, B, or C occurs.

2.4. Examples

Below, we will give some examples using the concepts learned above.

Example 1. Two dice are rolled. What is the probability of getting the following outcomes?

a) 5 and 6; b) the sum is 8.

Solution. a) Let A = {Get 5 and 6}. Name the two dice as Die 1 and Die 2.

$$\begin{aligned} &A \text{ occurs} \\ &= \{\text{Get 5 and 6}\} \\ &= \begin{cases} \textit{Die 1 gets 5 and Die 2 gets 6. These occur } \text{simultaneously.} \\ \quad \textit{or} \\ \textit{Die 1 gets 6 and Die 2 gets 5. These occur } \text{simultaneously.} \end{cases} \end{aligned}$$

So $$P(A) = P\{\text{Get 5 and 6}\} = \frac{\binom{1}{1}\binom{1}{1}}{\binom{6}{1}\binom{6}{1}} + \frac{\binom{1}{1}\binom{1}{1}}{\binom{6}{1}\binom{6}{1}} = \frac{1}{36} + \frac{1}{36} = \frac{1}{18}$$

b) Let $B = \{x + y = 8\} = \{x = i, y = 8 - i\}$

where $i = 2, 3, 4, 5, 6$

x is the number obtained by rolling Die 1.

y is the number obtained by rolling Die 2.

$$\text{B occurs} = \{x = i, y = 8 - i\} \text{ occurs} = \begin{cases} x = 2 \quad and \quad y = 6 \quad occur \quad \text{simultaneously.} \\ or \\ x = 3 \quad and \quad y = 5 \quad occur \quad \text{simultaneously.} \\ or \\ \vdots \\ or \\ x = 6 \quad and \quad y = 2 \quad occur \quad \text{simultaneously.} \end{cases}$$

So $$\mathrm{P(B)} = \mathrm{P}\{x + y = 8\} = \sum_{i=2}^{6} \mathrm{P}\{x = i\}\mathrm{P}\{y = 8 - i\}$$

Since all numbers on the die have an equal probability of occurrence

$$\mathrm{P}\{x = i\} = \mathrm{P}\{y = 8 - i\} = \frac{\binom{1}{1}}{\binom{6}{1}} = \frac{1}{6}$$

hence $$\mathrm{P(B)} = \sum_{i=2}^{6} \mathrm{P}\{x = i\}\mathrm{P}\{y = 8 - i\} = 5 \cdot \frac{1}{6} \cdot \frac{1}{6} = \frac{5}{36}$$

Example 2. What is the probability that a number drawn randomly from the following set of integers is divisible by 4?

(a) $\{1, 2, \ldots, 21\}$, (b) $\{1, 2, \ldots, 96\}$, (c) $\{1, 2, \ldots, 98\}$, (d) Z^+.

Solution.

(a) Let A = {The integer obtained is divisible by 4}. We can determine how many numbers are divisible by 4 from the given information.

That is, $$5 < \frac{21}{4} = 5.25$$

The left side of the inequality is 5, so among all the integers from 1 to 21, there are a total of 5 numbers that can be divided by 4, namely $\{4, 8, 12, 16, 20\}$.

So $\quad P(A) = \dfrac{\binom{5}{1}}{\binom{21}{1}} = \dfrac{5}{21}$

Because $\quad 5 < \dfrac{21}{4} \quad$ so $\quad \dfrac{5}{21} < \dfrac{1}{4}$

That is, $\quad P(A) = \dfrac{5}{21} < \dfrac{1}{4}$

(b) $\{1, 2, \ldots, 96\}$

Because $\quad \dfrac{96}{4} = 24 \quad$ so $\quad P(B) = \dfrac{\binom{24}{1}}{\binom{96}{1}} = \dfrac{24}{96} = \dfrac{1}{4}$

(c) $\{1, 2, \ldots, 98\}$

Because $\quad 24 < \dfrac{98}{4} \quad$ so $\quad P(C) = \dfrac{\binom{24}{1}}{\binom{98}{1}} = \dfrac{24}{98} = \dfrac{12}{49}$

(d) Since $Z^+ = \{1, 2, 3, \ldots, n\}$, there are n integers in this set.

$$k \le \frac{n}{4}$$

There are k integers out of n that are divisible by 4.

So $\quad P(D) = \dfrac{\binom{k}{1}}{\binom{n}{1}} = \dfrac{k}{n} \le \dfrac{1}{4}$

When $\dfrac{n}{4} = k$, that is, the remainder r of the division $\dfrac{n}{4}$ is r = 0,

$$P(D) = \frac{k}{n} = \frac{1}{4}$$

When $\frac{n}{4} > k$, that is, the remainder r of the division $\frac{n}{4}$ is $r \neq 0$,

$$P(D) = \frac{k}{n} < \frac{1}{4} \qquad \text{That is,} \quad P(D) \approx \frac{1}{4}$$

Example 3. A number is randomly selected from the integers 1 to 1000. What is the probability that the integer obtained is neither divisible by 4 nor divisible by 6?

Before solving this example, we first introduce the duality rule of sets.

$$\overline{A \cup B} = \overline{A} \cap \overline{B}$$

$$\overline{A \cap B} = \overline{A} \cup \overline{B}$$

The above formula can be generalised to n events.

$$\overline{\bigcup_{i=1}^{n} A_i} = \bigcap_{i=1}^{n} \overline{A}_i$$

$$\overline{\bigcap_{i=1}^{n} A_i} = \bigcup_{i=1}^{n} \overline{A}_i$$

Solution. Let the event A = {The integer obtained is divisible by 4}, event B = {The integer obtained is divisible by 6}.

Then

$\overline{A \cup B} = \overline{A} \cap \overline{B}$ = {The integer obtained is neither divisible by 4 nor divisible by 6}

So $\qquad P(\overline{A \cup B}) = 1 - P(A \cup B) = 1 - [P(A) + P(B) - P(A \cap B)]$

Because $\qquad \frac{1000}{4} = 250$

so $$P(A) = \frac{\binom{250}{1}}{\binom{1000}{1}} = \frac{1}{4}$$

Since $$166 < \frac{1000}{6} < 167$$

so $$P(B) = \frac{\binom{166}{1}}{\binom{1000}{1}} = \frac{83}{500}$$

When a number is divisible by 4 and 6 simultaneously, it is equivalent to being divisible by 12. Therefore

$$83 < \frac{1000}{12} < 84$$

We get $$P(A \cap B) = \frac{\binom{83}{1}}{\binom{1000}{1}} = \frac{83}{1000}$$

Hence, the probability that the integer obtained is neither divisible by 4 nor divisible by 6 is

$$p = P(\overline{A \cup B}) = 1 - (\frac{1}{4} + \frac{83}{500} - \frac{83}{1000}) = 0.667$$

That is, $p = 0.667$

Example 4. Nina took two books with her when she had a long train journey. The probability that she likes the first book is 0.6, and the probability that she likes the second book is 0.5. The probability that she likes both books is 0.4. What is the probability that she dislikes both books?

Solution. Given that $P(A) = P\{\text{She likes the first book}\} = 0.6$,

$P(B) = P\{\text{She likes the second book}\} = 0.5,$

$P(A \cap B) = \{\text{She likes both books}\} = 0.4$

then, by the duality rule, we get

$\overline{A \cup B} = \overline{A} \cap \overline{B}$

$= \{\text{Disliking the first book and disliking the second book occurred at the same time}\}$

$= \{\text{She dislikes both books}\}$

So

$$P(\overline{A \cup B}) = 1 - P(A \cup B) = 1 - [P(A) + P(B) - P(A \cap B)]$$
$$= 1 - (0.6 + 0.5 - 0.4) = 0.3$$

That is, the probability that she dislikes both books is

$$p = 0.3$$

Example 5. Two real numbers are randomly selected from [0, 3]. What is the probability that the difference between these two numbers is at most 2?

Solution. Let the two numbers chosen at random be x and y.

We know $\quad U = \{ (x, y) \mid 0 \le x \le 3,\ 0 \le y \le 3 \}$

$$A = \{ (x, y) \mid |y - x| \le 2,\ x \in [0, 3],\ y \in [0, 3] \}$$
$$= \{ (x, y) \mid x - 2 \le y \le x + 2,\ x \in [0, 3],\ y \in [0, 3] \}$$

Because all random points $Q_U(x, y)$ of U lie within the region of M(U) bounded by the straight lines $x = 0$, $x = 3$, $y = 0$, $y = 3$, as shown in Graph 2.5.

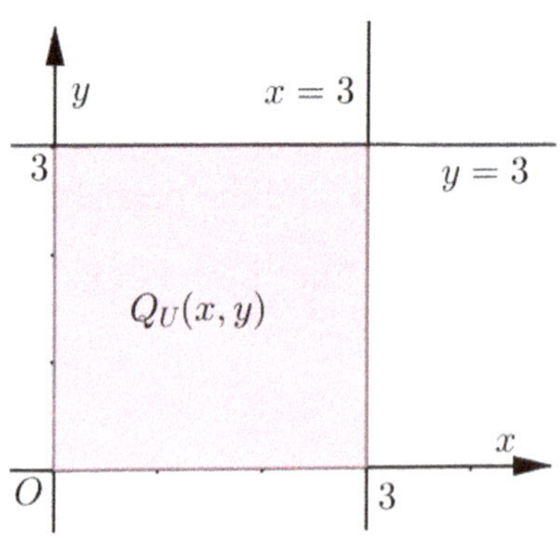

Graph 2.5

From Graph 2.5, we know that the square's area is $M(U) = 3 \times 3 = 9$.

Because all random points $Q_A(x, y)$ of A lie within the region M(A) bounded by the straight lines $y_1 = x - 2$, $y_2 = x + 2$, $x = 0$, $x = 3$, $y = 0$, $y = 3$, as shown in Graph 2.6.

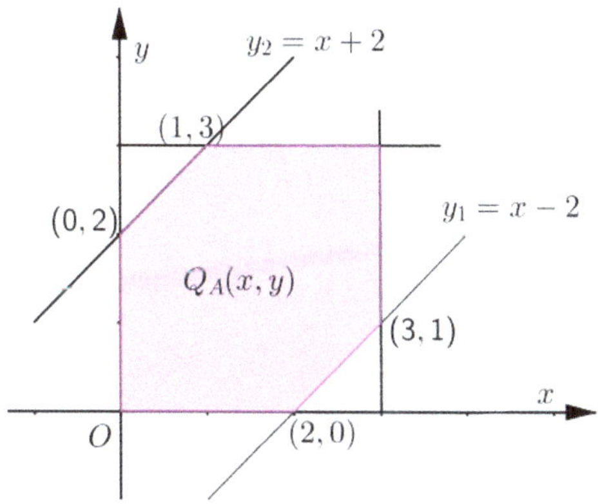

Graph 2.6

From Graph 2.6, we know the area of the region is

$$M(A) = M(U) - 2A_\Delta = 9 - 2\cdot \frac{1}{2} \cdot 1 \cdot 1 = 8$$

where A_Δ is the area of the triangle.

So $$P(A) = \frac{M(A)}{M(U)} = \frac{8}{9}$$

That is, two real numbers are randomly selected from [0, 3]. The probability that the difference between the two real numbers is at most 2 is $P(A) = \frac{8}{9}$.

Example 6. A coin is tossed 3 times. Find the probability of getting at least one head.

Solution. Let A = {A coin is tossed 3 times. Get at least one head.}

Because $U = \{0, H, HH, HHH\} = \{0\} \cup \{H, HH, HHH\}$

event $A = \{H, HH, HHH\}$, its complementary event $A' = \{0\}$

When a coin is tossed three times, and the number of obtaining heads is 0, tails are obtained three times. So

$A' = \{0\} = \{TTT\}$

Because $\quad$ A' occurs = {TTT} occurs simultaneously (all tosses are independent)

so $\quad P(A') = P\{TTT\} = P\{T\}P\{T\}P\{T\} = \frac{1}{2} \cdot \frac{1}{2} \cdot \frac{1}{2} = \frac{1}{8}$

Since $\quad P(U) = P(A \cup A') = P(A) + P(A')$

so $\quad P(A) = 1 - P(A') = 1 - \frac{1}{8} = \frac{7}{8}$

Example 7. Choose a card at random from a deck of playing cards.
Let A = {Get a picture card (Jack, Queen or King)}. B = {Get an ace}. Calculate P(A), P(B), and $P(A \cup B)$.

Solution.

$$P(A) = \frac{n(A)}{n(U)} = \frac{\binom{12}{1}}{\binom{52}{1}} = \frac{12}{52} = \frac{3}{13}$$

$$P(B) = \frac{n(B)}{n(U)} = \frac{\binom{4}{1}}{\binom{52}{1}} = \frac{4}{52} = \frac{1}{13}$$

Since $\quad A \cap B = \varnothing \quad$ so $\quad P(A \cup B) = P(A) + P(B) = \frac{3}{13} + \frac{1}{13} = \frac{4}{13}$

Example 8. Choose a card at random from a deck of playing cards. Let A = {Get an ace}, B = {not getting 2}. Are A and B independent?

Solution. Because $\quad n(B) = 52 - 4 = 48, \quad n(A) = 4$

so $$P(A|B) = \frac{P(A\cap B)}{P(B)} = \frac{n(A\cap B)}{n(B)} = \frac{n(A)}{n(B)} = \frac{\binom{4}{1}}{\binom{48}{1}} = \frac{4}{48} = \frac{1}{12}$$

Since $$P(A) = \frac{\binom{4}{1}}{\binom{52}{1}} = \frac{4}{52} = \frac{1}{13}$$ so $P(A|B) \neq P(A)$

Therefore, A and B are not independent.

Example 9. In 2000, 38.5% of Finns were married. What is the probability that a randomly selected Finnish person is a married woman?

Solution. Let the event A = {married}, B = {a randomly selected Finn is a married woman}.

Because the probability of being married is

$$P(A) = 38.5\% = \frac{38.5}{100} = \frac{\frac{38.5}{2}(women) + \frac{38.5}{2}(men)}{100}$$

In other words, there are $\frac{38.5}{2}$ married women among 100 people.

So $$P(B) = \frac{\binom{\frac{38.5}{2}}{1}}{\binom{100}{1}} = \frac{\frac{38.5}{2}}{100} = \frac{38.5}{200} = 0.1925$$

Example 10. What is the probability that a randomly selected day from the year 2001 is (1) a Sunday and a day in the summer (i.e., a day in June, July, or August), (2) a Sunday or a day in the summer?

(a) Find the approximate value. (b) Find the exact value.

Solution. (a) Let A = {Sunday}, B = {a day in the summer}.

Because there are 365 days a year and 7 days a week.

$$52 < \frac{365}{7}$$

That is, $n(\text{U}) = 365$ days $\quad n(\text{A}) \approx 52$ weeks.

So $$\text{P(A)} \approx \frac{\binom{52}{1}}{\binom{365}{1}} = \frac{52}{365}$$

On the other hand, there are 12 months in a year, so $n(\text{B}) = 3$ months.

So $$\text{P(B)} = \frac{\binom{3}{1}}{\binom{12}{1}} = \frac{3}{12} = \frac{1}{4}$$

Hence (1) $\text{P}(\text{A} \cap \text{B}) = \text{P(A)P(B)} \approx \frac{52}{365} \times \frac{1}{4} = \frac{13}{365} \approx 0.0356$

(2) $\text{P}(\text{A} \cup \text{B}) = \text{P(A)} + \text{P(B)} - \text{P}(\text{A} \cap \text{B}) \approx \frac{52}{365} + \frac{1}{4} - \frac{13}{365} = \frac{521}{1460} \approx 0.357$

(b) From the 2001 calendar we get $n(\text{A}) = 52$ (Sundays), $n(\text{B}) = 92$ summer days.

So $$\text{P(A)} = \frac{\binom{52}{1}}{\binom{365}{1}} = \frac{52}{365} \qquad \text{P(B)} = \frac{\binom{92}{1}}{\binom{365}{1}} = \frac{92}{365}$$

Since there are 13 Sundays out of 92 summer days. That is,

$$n(\text{A} \cap \text{B}) = 13$$

Hence (1) $$\text{P}(\text{A} \cap \text{B}) = \frac{\binom{13}{1}}{\binom{365}{1}} = \frac{13}{365} \approx 0.0356$$

(2) $P(A \cup B) = P(A) + P(B) - P(A \cap B) = \frac{52}{365} + \frac{92}{365} - \frac{13}{365} = \frac{131}{365} \approx 0.359$

Example 11. In a penalty kick, football player A has a probability of success of 0.85 and football player B has a probability of success of 0.7. During a penalty kick, A takes the penalty kick first, and B takes the second penalty kick. They each kick a ball. What is the probability that the following situation occurs?

They score (a) two goals, (b) 0 goal, (c) one goal.

Solution. Let A = {A succeeds}, B = {B succeeds}, A' = {A fails}, B' = {B fails}.

$$P(A) = 0.85 \quad P(B) = 0.7$$

$$P(A') = 1 - P(A) = 1 - 0.85 = 0.15$$

$$P(B') = 1 - P(B) = 1 - 0.7 = 0.3$$

(a) Event {two goals} occurs = A and B occur consecutively simultaneously.

So $\quad P\{\text{two goals}\} = P(A)P(B) = 0.85 \times 0.7 = 0.595$

(b) Event {0 goal} occurs = A' and B' occur consecutively simultaneously.

So $\quad P\{\text{0 goal}\} = P(A')P(B') = 0.15 \times 0.3 = 0.045$

(c)

$$\text{Event \{one goal\} occurs} = \begin{cases} \text{A} \quad \textit{and} \quad \text{B'} \quad \textit{occur} \quad \text{consecutively} \quad \text{simultaneously.} \\ \quad\quad \textit{or} \\ \text{B} \quad \textit{and} \quad \text{A'} \quad \textit{occur} \quad \text{consecutively} \quad \text{simultaneously.} \end{cases}$$

So $\quad P\{\text{one goal}\} = P(A, B') + P(B, A') = P(A)P(B') + P(B)P(A')$

$$= 0.85 \times 0.3 + 0.7 \times 0.15 = 0.36$$

Example 12. There are 5 red balls, 3 yellow balls and 2 white balls in a box. Take a ball out of the box and do not put it back. Take 3 balls in total. Find the probability of the following.

(a) The first two balls are red, and the third ball is white;

(b) All three balls are of the same colour.

Solution. (a) Let A = {The first two balls are red, and the third ball is white}.

So $$P(A)=\frac{\binom{5}{1}\binom{4}{1}\binom{2}{1}}{\binom{10}{1}\binom{9}{1}\binom{8}{1}}=\frac{1}{18}\approx 0.0556$$

(b) Let B = {All three balls are of the same colour}

Since there are only two white balls in the box, the three balls must be red or yellow. So

$$\text{event B occurs}=\begin{cases}\textit{Get red balls for 3 times. These occur simultaneously.}\\ \textit{or}\\ \textit{Get yellow balls for 3 times. These occur simultaneously.}\end{cases}$$

Hence $$P(B)=\frac{\binom{5}{1}\binom{4}{1}\binom{3}{1}}{\binom{10}{1}\binom{9}{1}\binom{8}{1}}+\frac{\binom{3}{1}\binom{2}{1}\binom{1}{1}}{\binom{10}{1}\binom{9}{1}\binom{8}{1}}=\frac{11}{120}\approx 0.0917$$

Example 13. A card is drawn at random from a deck of playing cards without replacement. Three cards are drawn in total in such a way. What is the probability that the first two cards are black spades and the third card is a red heart?

Solution. The sampling here is sampling without replacement.

Let A = {the first two cards are black spades, and the third card is a red heart}.

A deck of playing cards has 52 cards, including 13 black spades and 13 red hearts.

So $$P(A) = \frac{\binom{13}{1}\binom{12}{1}\binom{13}{1}}{\binom{52}{1}\binom{51}{1}\binom{50}{1}} = \frac{13}{850} \approx 0.0153$$

Example 14.

(a) Roll a die five times. What is the probability of getting the number 4 at least once?

(b) Roll two dice 30 times. What is the probability that both dice will give a 4 at least once?

Solution. Let A = {Get the number 4 in one throw}, its complementary event A' = {Do not get the number 4 in one throw}.

In one throw $n(\mathrm{U}) = \binom{6}{1}$ $n(\mathrm{A}) = \binom{1}{1}$

So $$P(A) = \frac{\binom{1}{1}}{\binom{6}{1}} = \frac{1}{6} \qquad P(A') = 1 - \frac{1}{6} = \frac{5}{6}$$

(a) Since the event {Do not get the number 4 in five throws} occurs

= A' occurs 5 times simultaneously

so P{Do not get 4 in five throws} = P{A', A', A', A', A'}

$$= P(A')P(A')P(A')P(A')P(A') = \left(\frac{5}{6}\right)^5$$

Therefore P{Roll a die five times. Get the number 4 at least once}

$$= 1 - P\{\text{Do not get 4 in five throws}\} = 1 - \left(\frac{5}{6}\right)^5 = \frac{4651}{7776} \approx 0.598$$

(b) Let B = {Roll two dice and both dice give a 4}.

So $$P(B) = \frac{\binom{1}{1}\binom{1}{1}}{\binom{6}{1}\binom{6}{1}} = \frac{1}{36}$$

B' is the complementary event of B.

B' = {Two dice are rolled simultaneously, and both dice do not give a 4 at the same time}

So $$P(B') = 1 - P(B) = 1 - \frac{1}{36} = \frac{35}{36}$$

Since the event {Two dice are rolled 30 times, and there is no single time when both dice give a 4 at the same time}

$$= \text{B' occurs 30 times simultaneously} = [P(B')]^{30} = \left(\frac{35}{36}\right)^{30}$$

so P{Roll two dice 30 times, at least once both dice give 4 simultaneously}

$$= 1 - \left(\frac{35}{36}\right)^{30} \approx 0.570$$

Example 15. Karol forgot the last digit of his friend's phone number. So he dialled randomly.

(1) Find the probability that he will get through the required call by dialling no more than 3 times.

(2) If it is known that the last number is odd, what is the probability?

Solution. (1) Let A = {Dial no more than 3 times to get through the desired call}.

$$
= \begin{cases} \textit{connected for the 1st try} \\ \textit{or} \\ \textit{unable to connect for the 1st try and connected for the 2nd try} \\ \textit{or} \\ \textit{unable to connect for the 1st and 2nd try and connected for the 3rd try} \end{cases}
$$

Here is sampling without replacement with 10 different numbers, so

$$
P(A) = \frac{\binom{1}{1}}{\binom{10}{1}} + \frac{\binom{9}{1}\binom{1}{1}}{\binom{10}{1}\binom{9}{1}} + \frac{\binom{9}{1}\binom{8}{1}\binom{1}{1}}{\binom{10}{1}\binom{9}{1}\binom{8}{1}} = 0.3
$$

(2) Let B = {It is known that the last number is an odd number}. Because there are 5 different odd numbers among the 10 numbers, and each time the number is randomly dialled among the 5 odd numbers, then under the condition of B, the probability of A occurring is

$$
P(A|B) = \frac{\binom{1}{1}}{\binom{5}{1}} + \frac{\binom{4}{1}\binom{1}{1}}{\binom{5}{1}\binom{4}{1}} + \frac{\binom{4}{1}\binom{3}{1}\binom{1}{1}}{\binom{5}{1}\binom{4}{1}\binom{3}{1}} = 0.6
$$

Method 2

(1) Because

P(A) + P{He dials 3 times without getting through to the required number} = 1

so P(A) = 1 – P{He dials 3 times without getting through to the required number}

$$
= 1 - \frac{9}{10} \cdot \frac{8}{9} \cdot \frac{7}{8} = 0.3
$$

(2) Similarly, we get

$$P(A|B) = 1 - \frac{4}{5}\cdot\frac{3}{4}\cdot\frac{2}{3} = 0.6$$

Example 16. The bag contains r red balls and t white balls. Take one ball from the bag at a time, observe its colour, and put it back into the bag, and put a more balls of the same colour as the ball taken out. If balls are taken from the bag 4 times in a row, what is the probability of getting a red ball for the first and second times and a white ball for the third and fourth times?

Solution. Let A_i ($i = 1, 2, 3, 4$) = {The red ball is obtained at the ith time}, then

$\bar{A}_3$ = {The white ball is obtained at the third time}

$\bar{A}_4$ = {The white ball is obtained at the fourth time}.

The required probability is

$$P(A_1A_2\bar{A}_3\bar{A}_4) = P(A_1)P(A_2|A_1)P(\bar{A}_3|A_1A_2)P(\bar{A}_4|A_1A_2\bar{A}_3)$$

$$= \frac{\binom{r}{1}}{\binom{r+t}{1}}\cdot\frac{\binom{r+a}{1}}{\binom{r+t+a}{1}}\cdot\frac{\binom{t}{1}}{\binom{r+t+2a}{1}}\cdot\frac{\binom{t+a}{1}}{\binom{r+t+3a}{1}}$$

$$= \frac{r}{r+t}\cdot\frac{r+a}{r+t+a}\cdot\frac{t}{r+t+2a}\cdot\frac{t+a}{r+t+3a}$$

Example 17. The probability that a glass bottle made by a glass bottle manufacturer breaks when it is dropped for the first time is $\frac{1}{3}$. If it does not break on the first drop, the probability of breaking on the second drop is $\frac{5}{9}$. If it is not broken in the first two falls, the probability of breaking in the third fall is $\frac{7}{8}$. Find the probability that the glass bottle does not break after being dropped three times.

Solution. Let A_i = {The glass bottle breaks when it falls for the *i*th time}, $i = 1, 2, 3$

B = {The glass bottle was dropped three times without breaking}.

Since $B = \bar{A}_1\bar{A}_2\bar{A}_3$

so $P(B) = P(\bar{A}_1\bar{A}_2\bar{A}_3) = P(\bar{A}_3|\bar{A}_1\bar{A}_2)P(\bar{A}_2|\bar{A}_1)P(\bar{A}_1)$

$$= (1-\frac{7}{8})(1-\frac{5}{9})(1-\frac{1}{3}) = \frac{1}{27}$$

That is, the probability that the glass bottle does not break after being dropped three times is $P(B) = \frac{1}{27}$.

Example 18. There is a system, as shown in Graph 2.7.

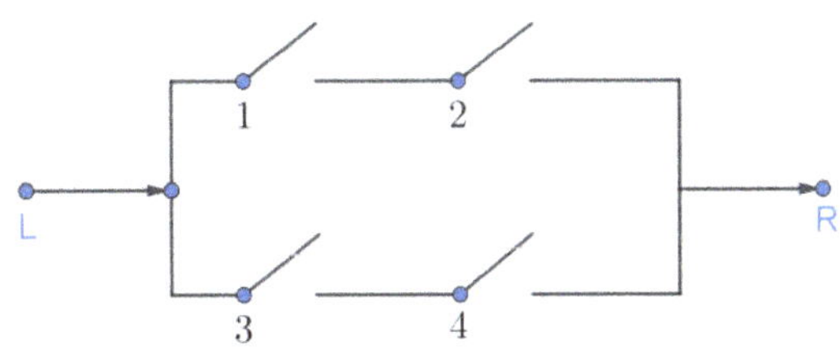

Graph 2.7

where 1, 2, 3, 4 are switches.

Let each switch be independent of each other, and the probability of each switch being closed is p. Find the probability that from L to R is a connected path.

Solution. Let A_i = {The *i*th switch is closed} $i = 1, 2, 3, 4$

B = {L to R is a connected path}.

Since B occurs $= \begin{cases} A_1 \quad \textit{and} \quad A_2 \quad \textit{occur} \quad \text{simultaneously.} \\ \qquad\qquad \textit{or} \\ A_3 \quad \textit{and} \quad A_4 \quad \textit{occur} \quad \text{simultaneously.} \\ \qquad\qquad \textit{or} \\ \quad A_1A_2A_3A_4 \quad \textit{occur} \quad \text{simultaneously.} \end{cases}$

That is, $B = A_1A_2 \cup A_3A_4$

Because A_1, A_2, A_3, A_4 are independent of each other, so

$$\begin{aligned} P(B) &= P(A_1A_2) + P(A_3A_4) - P(A_1A_2A_3A_4) \\ &= P(A_1)P(A_2) + P(A_3)P(A_4) - P(A_1)P(A_2)P(A_3)P(A_4) \\ &= p^2 + p^2 - p^4 \\ &= 2p^2 - p^4 \end{aligned}$$

That is, the probability that from L to R is a connected path is

$$P(B) = 2p^2 - p^4$$

Example 19. It is known that 6% of men are colour-blind, and 0.4% of women are colour-blind. Now, randomly select one person from a population with an equal number of males and females, and the person happens to be colour-blind. What is the probability that this person is a male?

Solution. Here, a person is selected randomly. This person is colour-blind, and this is certain. We need to find the probability that this person is male. Therefore, let the number of men and women be N, respectively. Then

Number of men who are colour-blind $= 6\% \cdot N$

Number of women who are colour-blind $= 0.4\% \cdot N$

Total number of colour-blind people in the population $= 6\%N + 0.4\%N = 6.4\%N$

So $\quad n(U) = \binom{6.4\%N}{1} \qquad n\{\text{The colour-blind person is a man}\} = \binom{6\%N}{1}$

Then

$$P\{\text{The colour-blind person is a man}\} = \frac{\binom{6\%N}{1}}{\binom{6.4\%N}{1}} = \frac{6\%N}{6.4\%N} = \frac{15}{16}$$

That is, $\quad P\{\text{The colour-blind person is a man}\} = \dfrac{15}{16}$

Example 20. In a population, 37.5% of people have blood type A, 20.9% have type B, 7.9% have type AB, and 33.7% have type O. The blood type pairs known to allow blood transfusions are shown in the following table. Now, choose one person randomly from the population to be the blood donor and then choose another person randomly to be the blood recipient. What is the probability of a successful blood transfusion?

blood donor \ blood recipient	blood type A	blood type B	blood type AB	blood type O
blood type A	✓	×	✓	✓
blood type B	×	✓	✓	✓
blood type AB	✓	✓	✓	✓
blood type O	×	×	×	✓

where ✓: Blood transfusion is allowed. ×: Blood transfusion is not allowed.

Solution. In the following, we first find the probability of successful pairing for each blood type.

Because

$$\{\text{successful pairing for type A}\}\text{ occurs} = \begin{cases} \text{A} \quad \textit{and} \quad \text{A} \quad \textit{occur} \quad \text{simultaneously.} \\ \qquad\qquad \textit{or} \\ \text{A} \quad \textit{and} \quad \text{AB} \quad \textit{occur} \quad \text{simultaneously.} \end{cases}$$

so
$$\begin{aligned} P\{\text{successful pairing for type A}\} &= P(A)P(A) + P(A)P(AB) \\ &= P(A)[P(A) + P(AB)] \end{aligned}$$

Similarly, we get

$$P\{\text{successful pairing for type B}\} = P(B)[P(B) + P(AB)]$$

$$P\{\text{successful pairing for type AB}\} = P(AB)[P(A) + P(B) + P(AB)]$$

$$P\{\text{successful pairing for type O}\} = P(O)[P(A) + P(B) + P(AB) + P(O)] = P(O) \cdot 1$$

Because {successful blood transfusion for randomly selected person} occurs =

$$\begin{cases} \{Type\quad \text{A}\quad successful\quad pairing\}\quad occurs \\ \qquad\qquad\qquad or \\ \{Type\quad \text{B}\quad successful\quad pairing\}\quad occurs \\ \qquad\qquad\qquad or \\ \{Type\quad \text{AB}\quad successful\quad pairing\}\quad occurs \\ \qquad\qquad\qquad or \\ \{Type\quad \text{O}\quad successful\quad pairing\}\quad occurs \end{cases}$$

Therefore

P{successful blood transfusion for randomly selected person}

= P(A)[P(A) + P(AB)] + P(B)[P(B) + P(AB)] + P(AB)[P(A) + P(B) + P(AB)] + P(O)

= 0.375(0.375 + 0.079) + 0.209(0.209 + 0.079) + 0.079(0.375 + 0.209 + 0.079)

+ 0.337

= 0.619819

That is,

P{successful blood transfusion for randomly selected person} = 0.619819

Example 21. Three persons, A, B, and C, shoot at a plane at the same time. The probability of their shots hitting the plane is 0.3, 0.6, and 0.8, respectively. The probability of the plane being hit and then shot down by one person is 0.1. The probability of the plane being hit and then shot down by two persons is 0.5. If all shots from three persons hit the plane, the plane will definitely be shot down. Find the probability that the plane is shot down.

Solution. Let (1) = {Shots from one person hit the plane},

(2) = {Shots from two persons hit the plane},

(3) = {Shots from three persons hit the plane},

G = {The plane is shot down}.

Because

$$(1) \text{ occurs} = \begin{cases} \text{A} \quad \textit{hitting} \quad \textit{and} \quad \text{B}, \quad \text{C} \quad \textit{not} \quad \textit{hitting} \quad \textit{occur} \quad \text{simultaneously} \\ \qquad\qquad\qquad\qquad \textit{or} \\ \text{B} \quad \textit{hitting} \quad \textit{and} \quad \text{A}, \quad \text{C} \quad \textit{not} \quad \textit{hitting} \quad \textit{occur} \quad \text{simultaneously} \\ \qquad\qquad\qquad\qquad \textit{or} \\ \text{C} \quad \textit{hitting} \quad \textit{and} \quad \text{A}, \quad \text{B} \quad \textit{not} \quad \textit{hitting} \quad \textit{occur} \quad \text{simultaneously} \end{cases}$$

so $\quad P(1) = 0.3 \cdot 0.4 \cdot 0.2 + 0.6 \cdot 0.7 \cdot 0.2 + 0.8 \cdot 0.7 \cdot 0.4 = 0.332$

Because

$$(2) \text{ occurs} = \begin{cases} \text{A}, \quad \text{B} \quad \textit{hitting} \quad \textit{and} \quad \text{C} \quad \textit{not} \quad \textit{hitting} \quad \textit{occur} \quad \text{simultaneously} \\ \qquad\qquad\qquad\qquad \textit{or} \\ \text{A}, \quad \text{C} \quad \textit{hitting} \quad \textit{and} \quad \text{B} \quad \textit{not} \quad \textit{hitting} \quad \textit{occur} \quad \text{simultaneously} \\ \qquad\qquad\qquad\qquad \textit{or} \\ \text{B}, \quad \text{C} \quad \textit{hitting} \quad \textit{and} \quad \text{A} \quad \textit{not} \quad \textit{hitting} \quad \textit{occur} \quad \text{simultaneously} \end{cases}$$

so $\quad P(2) = 0.3 \cdot 0.6 \cdot 0.2 + 0.3 \cdot 0.8 \cdot 0.4 + 0.6 \cdot 0.8 \cdot 0.7 = 0.468$

Because (3) occurs = A, B, C hitting occur simultaneously

so $\quad P(3) = 0.3 \cdot 0.6 \cdot 0.8 = 0.144$

For event G = {The plane is shot down}

$$\text{G occurs} = \begin{cases} (1) \quad \textit{and} \quad \textit{shoot-down} \quad \textit{occur} \quad \text{simultaneously} \\ \qquad\qquad\qquad \textit{or} \\ (2) \quad \textit{and} \quad \textit{shoot-down} \quad \textit{occur} \quad \text{simultaneously} \\ \qquad\qquad\qquad \textit{or} \\ (3) \quad \textit{and} \quad \textit{shoot-down} \quad \textit{occur} \quad \text{simultaneously} \end{cases}$$

Therefore, the probability that the plane is shot down is

$$\begin{aligned} P(G) &= P(1) \cdot 0.1 + P(2) \cdot 0.5 + P(3) \cdot 1 \\ &= 0.332 \cdot 0.1 + 0.468 \cdot 0.5 + 0.144 \cdot 1 = 0.4112 \end{aligned}$$

That is, $\quad P(G) = 0.4112$

From the above examples, we know that to determine the occurrence probability of an event, we need to analyse the event's meaning or the relationship between events. Set theory shows only the symbolic and operational significance of events. We need to be clear about this.

Exercise 2.4

1. A box contains 5 white balls, 3 red balls and 4 black balls. Take at random 3 balls from the box. Find the probability that among the 3 balls taken

- **a.** there is no red ball;
- **b.** there is exactly one red ball;
- **c.** there is at least one red ball;
- **d.** they are of the same colour;
- **e.** they are of a different colour.

2. Max has 5 keys, but only one can open the door. But he forgot which one it was, so he had to try them one by one. Find the probability of the following events.

- **a.** He managed to open the door on his first try.
- **b.** He managed to open the door on his first two attempts.
- **c.** He managed to open the door after no more than 5 attempts.
- **d.** He tried 5 times but failed to open the door.

3. Choose a number randomly from the natural numbers 1 to 100. Calculate the probability that the number is divisible by 10 or 11.

4. There are 5 line segments with lengths 1, 3, 5, 7, and 9 respectively. Choose 3 line segments at random. What is the probability of forming a triangle?

5. There are $2n$ ($n > 1$) equidistant points on the circumference of a circle. Three points are randomly selected to form a triangle. What is the probability of forming a right-angled triangle?

6. An elevator in a 20-story building picks up 8 passengers on the ground floor (first floor). The elevator stops at every floor. It is equally possible for every passenger to leave on each floor. Find the probability that no two passengers leave on the same floor.

7. There are 12 parking spaces in a car park. What is the probability that after 8 cars are parked, there will be 4 consecutive empty spaces in a row?

8. A and B agreed to meet at a coffee shop between 10 a.m. and 11 a.m. The person who arrives first waits for the other person for 20 minutes and then leaves. Find the probability that these two people can meet.

9. A total of 50 applications for summer jobs were received from three regions. There were 10, 15, and 25 applications from the three regions, respectively, with 3, 7, and 5 applications from girls. One region is randomly selected, and two applications are taken from that region, one after the other.

- **a.** Find the probability that the first one drawn is a girl's application.
- **b.** Given that the one drawn the last is a boy's application, find the probability that the first one drawn is a girl's application.

10. Toss a fair die 12 times. Find the probability that the numbers 1, 2, 3, 4, 5, and 6 all appear twice.

11. There are 8 classes in the first grade of a school, 6 classes in the second grade, and 4 classes in the third grade. All classes were randomly divided into three groups, with the same number of classes in each group. Find the probability that each group has class(es) in the third grade.

12. Six differently numbered balls, a, b, c, d, e, and f, are placed into six boxes numbered 1, 2, 3, 4, 5, and 6, respectively. Find the probability that ball a is not put into box 1 and ball b is not placed into box 6.

13. Consider a quadratic equation $x^2 + Bx + C = 0$, where B and C are the number of points appearing one after another when a fair die is thrown twice. Find the probability that

a. the equation has repeated roots;

b. the equation has real roots.

14. Throw a die continuously until the number 6 appears. Find the probability that an even number of times has been thrown eventually.

15. To prevent accidents, two alarm devices, A and B, are installed in the corridor of an office building. It is known that the probability that alarm device A works when it is used alone is 0.95, and the probability that alarm device B works when it is used alone is 0.9. The probability that alarm device B does not work when alarm device A does not work is 0.86. Find the probability that at least one alarm device works when an accident occurs.

16. Roll 10 dice at the same time. Repeat 5 times. Find the probability that all the dice will show the number 1 at least once.

17. Find the probability that the following systems can work normally. A letter within a rectangle represents a component. Components A, B, C, and D work with probability p_A, p_B, p_C, p_D, respectively.

a.

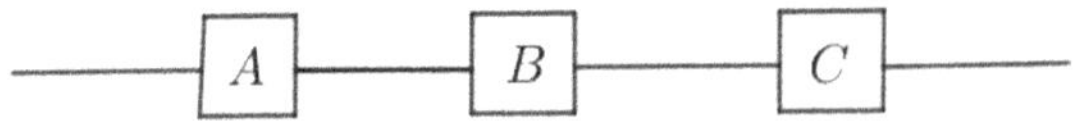

b.

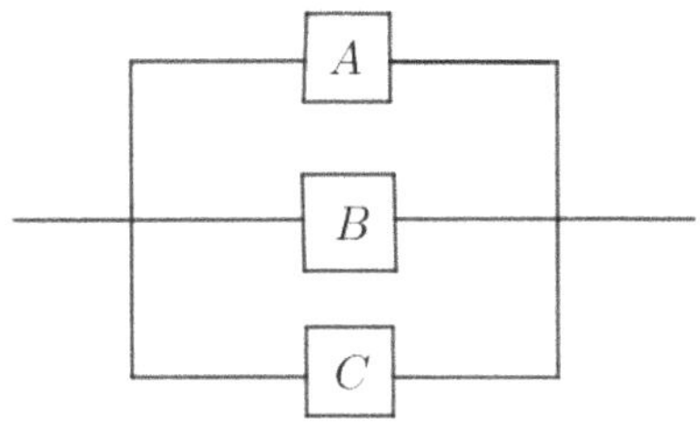

c.

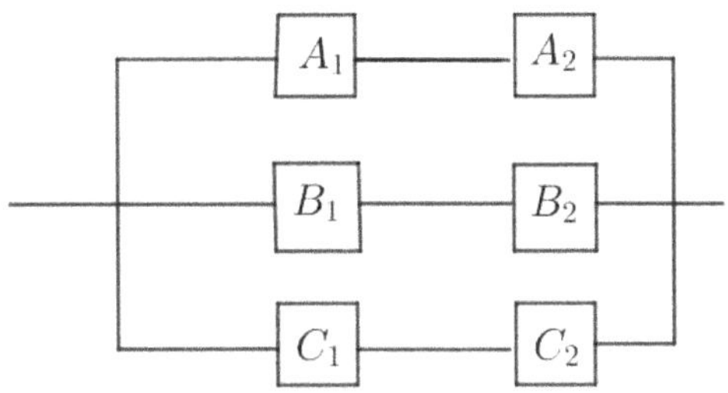

d.

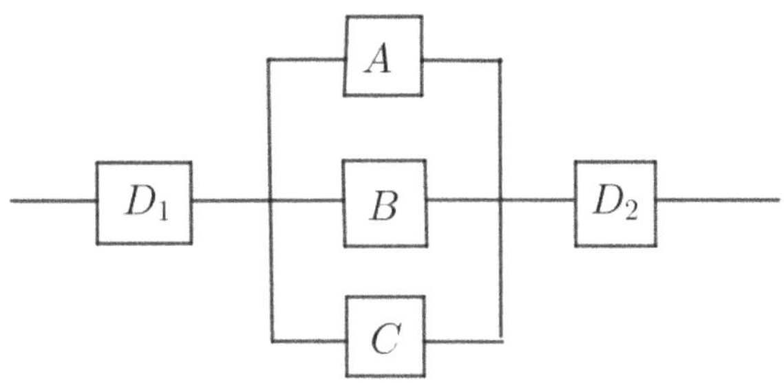

2.5. Law of total probability and Bayes' Theorem

In the following, we derive two important formulae for calculating probability. First, we introduce the definition of a sample space partition.

Definition

Let U be the sample space of an experiment. $B_1, B_2, \ldots, B_n$ are events of the experiment.

If (i) $B_iB_j = \varnothing, \quad i \neq j, \quad i, j = 1, 2, \ldots, n$;

(ii) $B_1 \cup B_2 \cup \ldots \cup B_n = U$

(iii) $1 = P(U) = P(B_1 \cup B_2 \cup \ldots \cup B_n) = P(B_1) + P(B_2) + \ldots + P(B_n) = \sum_{j=1}^{n} P(B_j)$

that is, $\sum_{j=1}^{n} P(B_j) = 1$

then we say that $B_1, B_2, \ldots, B_n$ is a partition of the sample space U.

U

B_1	B_2	…	…
…	…	…	…
…	…	…	…
…	…	B_{n-1}	B_n

If $B_1, B_2, \ldots, B_n$ is a partition of the sample space, then in each experiment, one and only one of the events $B_1, B_2, \ldots, B_n$ will definitely occur.

For example, let an experiment be "tossing a die and observe the number obtained". Its sample space is $U = \{1, 2, 3, 4, 5, 6\}$. A set of events of the experiment $B_1 = \{1, 2, 3\}$, $B_2 = \{4, 5\}$, $B_3 = \{6\}$ is a partition of the sample space U, while the set of events $C_1 = \{1, 2, 3\}$, $C_2 = \{3, 4\}$, $C_3 = \{5, 6\}$ is not a partition of U.

Theorem

If the sample space of an experiment is U, A is an event of the experiment, $B_1, B_2, \ldots, B_n$ is a partition of the sample space U and $P(B_j) > 0$ ($j = 1, 2, \ldots, n$), $\sum_{j=1}^{n} P(B_j) = 1$,

then we have

$$P(A) = P(A \mid B_1)P(B_1) + P(A \mid B_2)P(B_2) + \dots + P(A \mid B_n)P(B_n) = \sum_{j=1}^{n} P(A|B_j)P(B_j)$$

This is called the **law of total probability**.

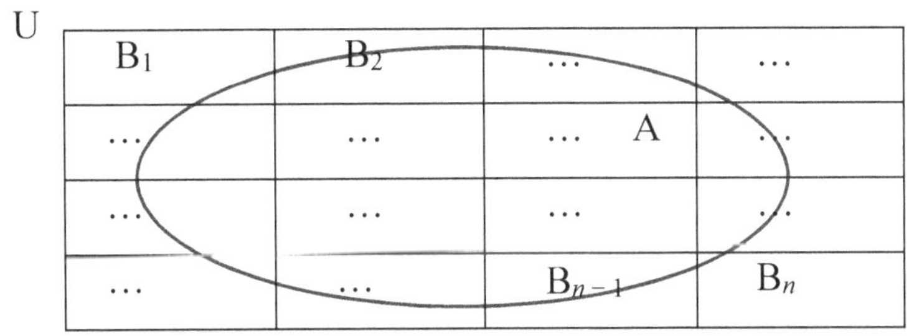

In many practical problems, it is not easy to calculate P(A), but it is easy to find a partition of U — $B_1, B_2, \dots, B_n$ and $P(B_i)$ and $P(A \mid B_i)P(B_i)$ are either known or easy to get, then we can use the law of total probability to find P(A).

Proof. Because

$$A = AU = A(B_1 \cup B_2 \cup \dots \cup B_n) = AB_1 \cup AB_2 \cup \dots \cup AB_n$$

From the assumption $P(B_i) > 0$ ($i = 1, 2, \dots, n$) and $(AB_i)(AB_j) = \varnothing$, $i \neq j$,

$\sum_{j=1}^{n} P(B_j) = 1$,we get

$$\begin{aligned} P(A) &= P(AB_1) + P(AB_2) + \dots + P(AB_n) \\ &= P(A \mid B_1)P(B_1) + P(A \mid B_2)P(B_2) + \dots + P(A \mid B_n)P(B_n) \\ &= \sum_{j=1}^{n} P(A|B_j)P(B_j) \end{aligned}$$ □

*In particular, the law of total probability also applies to conditional probability P(A|C). The formula is as follows.

$$P(A|C) = P(A|B_1C)P(B_1|C) + P(A|B_2C)P(B_2|C) + \dots + P(A|B_nC)P(B_n|C)$$

Another important formula is **Bayes' Theorem.**

Bayes' Theorem

If the sample space of an experiment is U, A is an event of the experiment, $B_1, B_2, \ldots, B_n$ is a partition of the sample space U and $P(B_i) > 0$ ($i = 1, 2, \ldots, n$), then we have

$$P(B_i \mid A) = \frac{P(A \mid B_i)P(B_i)}{\sum_{j=1}^{n} P(A \mid B_j)P(B_j)} \qquad i = 1, 2, \ldots, n$$

Proof 1. From the definition of conditional probability and the law of total probability, we have

$$P(B_i \mid A) = \frac{P(B_i A)}{P(A)} = \frac{P(A \mid B_i)P(B_i)}{\sum_{j=1}^{n} P(A \mid B_j)P(B_j)} \qquad i = 1, 2, \ldots, n$$

Proof 2. From the multiplication principle and the law of total probability, we have

$$P(AB_i) = P(B_i \mid A)P(A) = P(A \mid B_i)P(B_i)$$

That is, $P(B_i \mid A)P(A) = P(A \mid B_i)P(B_i)$

So
$$P(B_i \mid A) = \frac{P(B_i A)}{P(A)} = \frac{P(A \mid B_i)P(B_i)}{\sum_{j=1}^{n} P(A \mid B_j)P(B_j)} \qquad i = 1, 2, \ldots, n$$

Example 1. Three silicon wafer manufacturing companies supply the silicon wafers used in an electronic factory. According to previous data, we have the following records

Manufacturing company	rate of defective	quota in providing silicon wafers
1	0.02	0.12
2	0.01	0.82
3	0.03	0.06

The silicon wafers from these three companies are well-mixed and stored in the warehouse. There are no signs to tell the difference.

(1) One silicon wafer is taken randomly from the warehouse. Find the probability that it is defective.

(2) One silicon wafer is taken randomly from the warehouse. Suppose we know the selected silicon wafer is defective, and we need to find out which manufacturing company it comes from. In that case, we need to find out the probability that it comes from each manufacturing company. Find these probabilities.

Solution. Let A = {the selected one is defective}, B_i ($i = 1, 2, 3$) = {the selected one is supplied by the B_i manufacturing company}. It is easy to know that B_1, B_2, B_3 is a partition of the sample space U.

$$P(B_1) = 0.12, \quad P(B_2) = 0.82, \quad P(B_3) = 0.06,$$

$$P(A \mid B_1) = 0.02, \quad P(A \mid B_2) = 0.01, \quad P(A \mid B_3) = 0.03$$

(1) From the law of total probability

$$P(A) = P(A \mid B_1)P(B_1) + P(A \mid B_2)P(B_2) + P(A \mid B_3)P(B_3)$$

$$= 0.02 \times 0.12 + 0.01 \times 0.82 + 0.03 \times 0.06 = 0.0124$$

(2) From Bayes' Theorem

$$P(B_1 \mid A) = \frac{P(A|B_1)P(B_1)}{P(A)} = \frac{0.02 \times 0.12}{0.0124} \approx 0.194$$

$$P(B_2 \mid A) = \frac{P(A|B_2)P(B_2)}{P(A)} = \frac{0.01 \times 0.82}{0.0124} \approx 0.661$$

Since $B_1 \cup B_2 \cup B_3 = U \quad B_i \cap B_j = \varnothing \quad i \neq j \quad i, j = 1, 2, 3$

so $P(B_3 \mid A) = 1 - P(B_1 \mid A) - P(B_2 \mid A) = 1 - 0.194 - 0.661 = 0.145$

These results show that the probability that the defective ones came from the second company is the highest.

Example 2. The analysis of past data shows that when a machine is well-adjusted, the probability that the products produced by the machine are defect-free is 95%. When something goes wrong with the machine, there is a 20% chance that the products will not be defective. When the machine starts running every morning, the probability that it is well-adjusted is 80%. One morning, if the first product has no defect, find the probability that the machine is well-adjusted.

Solution. Let A = {the product is not defective}, B = {the machine is well adjusted}.

$$P(A\,|\,B) = 0.95,\ P(A\,|\,B') = 0.2,\quad P(B) = 0.8,\quad P(B') = 0.2$$

B, B' is a partition of the sample space U. From Bayes' Theorem

$$P(B\,|\,A) = \frac{P(A|B)P(B)}{P(A|B)P(B) + P(A|B')P(B')} = \frac{0.95\times 0.8}{0.95\times 0.8 + 0.2\times 0.2} = 0.95$$

Example 3. A student may know the correct answer or guess one when answering a multiple-choice question. Let p be the probability that he knows the correct answer. Then, $1-p$ represents the probability of guessing. If the student guesses, the probability of getting the correct answer is $\frac{1}{m}$. Here, m is the number of possible answers to the multiple-choice question. Find the probability that the student knows the correct answer, given that he has answered correctly.

Solution. Let A = {The student answered correctly},

B = {The student knows the correct answer},

B' = {The student does not know the correct answer}.

Then $\quad P(B) = p \qquad P(B') = 1-p \qquad P(A|\,B') = \frac{1}{m} \qquad P(A\cap B) = p$

B, B' is a partition of the sample space U. So according to Bayes' Theorem

$$P(B|A)=\frac{P(A\cap B)}{P(A)}=\frac{P(A\cap B)}{P(A\cap B)+P(A|B')P(B')}=\frac{p}{p+\frac{1}{m}(1-p)}=\frac{mp}{1+(m-1)p}$$

That is, under the condition that the answer is correct, the probability that the student knows the correct answer is

$$P(B|A)=\frac{mp}{1+(m-1)p}$$

For example, $m=3, p=\frac{2}{3}$. Then, given that the student's answer is known to be correct, the probability that he knows the correct answer is $\frac{6}{7}$.

Example 4. In a particular month, the probabilities of good weather, rain or snow are $\frac{1}{2}$, $\frac{1}{3}$ and $\frac{1}{6}$, respectively. Under different weather conditions, the probability of a student being late for school is $\frac{1}{4}$, $\frac{2}{5}$ and $\frac{5}{8}$, respectively. Find

(1) on any day, the probability that the student arrives at school on time.

(2) the probability of snow if the student is late.

Solution. Let A = {good weather}, B = {rain}, C ={snow}, L = {late for school}. It is given $P(A)=\frac{1}{2}$, $P(B)=\frac{1}{3}$, $P(C)=\frac{1}{6}$. A, B, C is a partition of the sample space U.

$$P(L|A)=\frac{1}{4},\ P(L|B)=\frac{2}{5},\ P(L|C)=\frac{5}{8}$$

(1) Since

L = {Late for school}

$$= \begin{cases} \textit{Good weather and being late for school occur } \text{simultaneously} \\ \textit{or} \\ \textit{Raining and being late for school occur } \text{simultaneously} \\ \textit{or} \\ \textit{Snowing and being late for school occur } \text{simultaneously} \end{cases}$$

so $$P(L) = P(L|A)P(A) + P(L|B)P(B) + P(L|C)P(C)$$

$$= \frac{1}{4} \cdot \frac{1}{2} + \frac{2}{5} \cdot \frac{1}{3} + \frac{5}{8} \cdot \frac{1}{6} = \frac{29}{80}$$

Then, the probability that the student arrives at school on time is

$$P\{\text{Arrive at school on time}\} = 1 - P(L) = 1 - \frac{29}{80} = \frac{51}{80} = 0.6375$$

That is, $$P\{\text{Arrive at school on time}\} = 0.6375$$

(2) If the student is late, the probability of snow is

$$P(C|L) = \frac{P(C \cap L)}{P(L)} = \frac{P(L|C)P(C)}{P(L)} = \frac{\frac{5}{8} \cdot \frac{1}{6}}{\frac{29}{80}} = \frac{25}{87} \approx 0.287$$

That is, $$P(C|L) \approx 0.287$$

Example 5. During a criminal investigation, the investigator was 80% confident that the suspect was guilty. A new piece of evidence suggests the criminal has specific physical characteristics (left-handedness, baldness, brown hair, etc.). If 30% of the people in the crowd have this characteristic, how confident is the prosecutor that the suspect is indeed guilty of this crime if he has this characteristic?

Solution. Let A = {The suspect is indeed guilty of the crime}, B = {The suspect has the physical characteristics of the criminal}.

Because $P(A) = 0.8$, $P(A') = 0.2$

A and A' is a partition of the sample space U.

$$P(B|A) = 1, \qquad P(B|A') = 0.3$$

According to Bayes' formula, we have

$$P(A|B) = \frac{P(A\cap B)}{P(B)} = \frac{P(B|A)P(A)}{P(B|A)P(A)+P(B|A')P(A')} = \frac{1\times 0.8}{1\times 0.8+0.3\times 0.2} \approx 0.930$$

That is, the prosecutor's confidence that the suspect committed the crime is

$$P(A|B) \approx 0.930$$

Example 6. An aircraft is missing at sea. It is estimated that it crashed in 4 regions with equal probability. Let $1-\gamma_i$ represent the probability that the plane crashed in the ith region is found. γ_i is known as the probability of oversight (i.e., failure to detect) and it is determined by the geographic location and environmental conditions of the region. It is known that a search of region 1 did not reveal any aircraft. Under this condition, find the conditional probability that the plane crashes in the ith region (i = 1, 2, 3, 4).

Solution. Let A_i = {The aircraft crashed in the ith region}, i = 1, 2, 3, 4.

B = {a search of region 1 did not reveal any aircraft}.

Because $P(A_i) = \frac{1}{4}$. That is, A_1, A_2, A_3, A_4 is a partition of the sample space U.

$$P(B|A_1) = \gamma_1, \qquad P(B|A_2) = P(B|A_3) = P(B|A_4) = 1$$

According to Bayes' formula, we have

$$P(A_1|B) = \frac{P(A_1\cap B)}{P(B)} = \frac{P(B|A_1)P(A_1)}{\sum_{i=1}^{4} P(B|A_i)P(A_i)} = \frac{\gamma_1\times\frac{1}{4}}{\gamma_1\times\frac{1}{4}+1\times\frac{1}{4}+1\times\frac{1}{4}+1\times\frac{1}{4}} = \frac{\gamma_1}{\gamma_1+3}$$

For j = 2, 3, 4 we have

$$P(A_j|B) = \frac{P(B|A_j)P(A_j)}{\sum_{i=1}^{4} P(B|A_i)P(A_i)} = \frac{1\times\frac{1}{4}}{\gamma_1\times\frac{1}{4}+1\times\frac{1}{4}+1\times\frac{1}{4}+1\times\frac{1}{4}} = \frac{1}{\gamma_1+3} \qquad j = 2, 3, 4$$

That is, $P(A_1|B) = \frac{\gamma_1}{\gamma_1+3}$, $P(A_2|B) = P(A_3|B) = P(A_4|B) = \frac{1}{\gamma_1+3}$

Example 7. Insurance companies believe that people who buy car insurance can be divided into two categories: those prone to accidents and those who are safe. Their statistics show that the probability of an accident-prone person getting into an accident within a year after buying car insurance is 0.5. The probability of a safe person getting into an accident within a year after purchasing car insurance is 0.1. 20% of the population is accident-prone. Now a new person is coming to buy car insurance.

(1) Find the probability that this person will have an accident within one year after buying car insurance.

(2) Find the probability that he is accident-prone if an accident occurs within a year.

(3) * If the person has already had an accident in the first year, find the probability of having another accident in the second year.

Solution. (1)

Let A = {People who have an accident within one year after buying the insurance}

B = {People who are prone to accidents in the population}

then B' = {People who are safe in the population}.

It is given $P(A|B) = 0.5$, $P(A|B') = 0.1$, $P(B) = 0.2$, $P(B') = 0.8$.

B and B' is a partition of the sample space U. Therefore, according to the total probability formula, we have

$$P(A) = P(A|B)P(B) + P(A|B')P(B') = 0.5 \cdot 0.2 + 0.1 \cdot 0.8 = 0.18$$

That is, the probability that this person will have an accident within one year after buying insurance is $P(A) = 0.18$.

(2) The required probability is P(B|A). According to Bayes' formula, we get

$$P(B|A) = \frac{P(B \cap A)}{P(A)} = \frac{P(A|B)P(B)}{P(A)} = \frac{0.5 \cdot 0.2}{0.18} = \frac{5}{9}$$

That is, the probability that he is an accident prone person is

$$P(B|A) = \frac{5}{9}$$

(3)* Let C = {another accident in the second year}. Then, according to the total probability formula, we have

$$P(C|A) = P(C|AB)P(B|A) + P(C|AB')P(B'|A)$$

Because $\quad P(B|A) = \frac{5}{9} \qquad P(B'|A) = 1 - P(B|A) = \frac{4}{9}$

$$P(C|AB) = P(C|B) = 0.5 \qquad P(C|AB') = P(C|B') = 0.1$$

Here, A and C are conditionally independent with respect to B or B'.

So, the required probability is

$$P(C|A) = 0.5 \times \frac{5}{9} + 0.1 \times \frac{4}{9} = \frac{29}{90}$$

That is, $\quad P(C|A) = \frac{29}{90}$

Example 8. Based on past clinical records, a specific test for diagnosing cancer has the following effects. If let A = {the test reaction is positive}, B = {the person diagnosed has cancer}, then we have P(A|B) = 0.98, P(A'|B') = 0.97. Now, a random experiment on the population will be conducted. Let the probability that the person being tested have cancer be 0.008. That is, P(B) = 0.008. Find P(B|A).

Solution. It is given that

$$P(A|B) = 0.98, \qquad P(A|B') = 1 - P(A'|B') = 1 - 0.97 = 0.03$$

$$P(B) = 0.008, \qquad P(B') = 0.992$$

B, B' is a partition of the sample space U. According to Bayes' formula, we get

$$P(B|A) = \frac{P(A|B)P(B)}{P(A|B)P(B) + P(A|B')P(B')} = \frac{0.98 \cdot 0.008}{0.98 \cdot 0.008 + 0.03 \cdot 0.992} \approx 0.209$$

That is, $P(B|A) \approx 0.209$

The results here show that although P(A|B) = 0.98 and P(A'|B') = 0.97, both probabilities are relatively high. However, if the test is used in a census, then P(B|A) ≈ 0.209. That is, out of 1,000 people with a positive reaction, only 209 will be diagnosed with cancer. If this is not noted, a wrong diagnosis will be made. This also shows that confusing P(A|B) with P(B|A) will cause adverse consequences.

Example 9. A batch of 100 calculators will be inspected and accepted by a school. The acceptance plan is as follows: Randomly select 3 calculators from this batch for testing (Let the tests of the 3 calculators be independent of each other). If at least one of the 3 calculators is found to be defective during testing, the batch of calculators will be rejected. Let the probability of a defective calculator being detected in a test be 0.95 and the probability of a non-defective calculator being detected in a test be 0.99. If it is known that 4 of these 100 calculators are defective, what is the probability that this batch of calculators will be accepted?

Solution.

Let B_i = {Randomly select 3 calculators and there are i defective calculators}

$i = 0, 1, 2, 3$

B_0, B_1, B_2, B_3 is a partition of the sample space U.

A = {The calculators are accepted}.

It is known that the probability of a non-defective calculator being detected in the test is 0.99. The probability that a defective calculator is mistakenly considered non-defective during testing is 1 – 0.95 = 0.05. And the tests of the calculators are independent of each other.

Therefore, under the condition that B_i has occurred, the probabilities of accepting this batch of calculators are respectively

$$P(A|B_0) = (0.99)^3 \qquad P(A|B_1) = (0.99)^2 \cdot 0.05$$

$$P(A|B_2) = 0.99 \cdot (0.05)^2 \qquad P(A|B_3) = (0.05)^3$$

And

$$P(B_0) = \frac{\binom{96}{3}}{\binom{100}{3}} \qquad P(B_1) = \frac{\binom{4}{1}\binom{96}{2}}{\binom{100}{3}}$$

$$P(B_2) = \frac{\binom{4}{2}\binom{96}{1}}{\binom{100}{3}} \qquad P(B_3) = \frac{\binom{4}{3}}{\binom{100}{3}}$$

Therefore

$$P(A) = \sum_{i=0}^{3} P(A|B_i)P(B_i)$$

$$= (0.99)^3 \cdot \frac{\binom{96}{3}}{\binom{100}{3}} + (0.99)^2 \cdot 0.05 \cdot \frac{\binom{4}{1}\binom{96}{2}}{\binom{100}{3}} + 0.99 \cdot (0.05)^2 \cdot \frac{\binom{4}{2}\binom{96}{1}}{\binom{100}{3}}$$

$$+ (0.05)^3 \cdot \frac{\binom{4}{3}}{\binom{100}{3}}$$

$$\approx 0.863$$

That is, the probability that this batch of calculators is accepted is

$$P(A) \approx 0.863$$

Example 10. According to the data analysis of records, there are three situations in which an item transported by a truck is damaged: 3% damage (this event is denoted as A_1), 8% damage (event A_2), and 80% damage (event A_3). It is known that $P(A_1) = 0.75$, $P(A_2) = 0.18$, $P(A_3) = 0.07$. Now, randomly pick 3 items from the items that have been transported and find that these three items are all good (this event is denoted as B). Find $P(A_1|B)$, $P(A_2|B)$, $P(A_3|B)$. (There are a lot of items here. Taking out one item does not affect the probability of whether the next one is a good item.)

Solution. Because $B = \{\text{The 3 items taken out are all good}\}$. There are three situations when an item is good.

(1) In event A_1, $P\{\text{good}\} = 1 - 3\% = 97\%$

(2) In event A_2, $P\{\text{good}\} = 1 - 8\% = 92\%$

(3) In event A_3, $P\{\text{good}\} = 1 - 80\% = 20\%$

So $\quad P(B|A_1) = 0.97^3, \quad P(B|A_2) = 0.92^3, \quad P(B|A_3) = 0.2^3$

It is given $\quad P(A_1) = 0.75, \quad P(A_2) = 0.18, \quad P(A_3) = 0.07.$

A_1, A_2, A_3 is a partition of the sample space U. According to Bayes' formula, we get

$$P(A_1|B) = \frac{P(BA_1)}{P(B)} = \frac{P(BA_1)}{P(BA_1) + P(BA_2) + P(BA_3)}$$

$$= \frac{0.97^3 \cdot 0.75}{0.97^3 \cdot 0.75 + 0.92^3 \cdot 0.18 + 0.2^3 \cdot 0.07} \approx 0.829$$

Similarly, we get

$$P(A_2|B) = \frac{0.92^3 \cdot 0.18}{0.97^3 \cdot 0.75 + 0.92^3 \cdot 0.18 + 0.2^3 \cdot 0.07} \approx 0.170$$

$$P(A_3|B) = \frac{0.2^3 \cdot 0.07}{0.97^3 \cdot 0.75 + 0.92^3 \cdot 0.18 + 0.2^3 \cdot 0.07} \approx 0.000679$$

Example 11*. (Laplace's rule of succession) The box has $k + 1$ (numbered from 0 to k) uneven coins. When tossing the ith coin, the probability of heads appearing is $\frac{i}{k}$, $i = 0, 1, \ldots, k$. Take a coin randomly from the box and toss it repeatedly. If the first n tosses all result in heads, find the probability that the $(n + 1)$th result still comes out as head.

Solution. Let B_i = {The coin taken randomly is the ith coin}.

Its probability distribution is $\mathrm{P}(B_i) = \frac{1}{k+1}$, $\sum_{i=0}^{k} \mathrm{P}(B_i) = 1$.

$B_0, B_1, \ldots, B_k$ is a partition of the sample space U.

Let F_n = {The first n tosses result in heads}, H = {The $(n + 1)$th toss comes up head}.

Then, according to the total probability formula, we have

$$\mathrm{P}(\mathrm{H}|F_n) = \sum_{i=0}^{k} \mathrm{P}(\mathrm{H}|F_n B_i)\mathrm{P}(B_i|F_n)$$

Because the results of each toss are conditionally independent, the probability of a head coming up each time is $p = \frac{i}{k}$. Therefore, we have

$$\mathrm{P}(\mathrm{H}|F_n B_i) = \mathrm{P}(\mathrm{H}|B_i) = \frac{i}{k}$$

$$\mathrm{P}(F_n|B_i) = \left(\frac{i}{k}\right)^n$$

$$\mathrm{P}(B_i|F_n) = \frac{\mathrm{P}(B_i F_n)}{\mathrm{P}(F_n)} = \frac{\mathrm{P}(F_n|B_i)\mathrm{P}(B_i)}{\sum_{j=0}^{k} \mathrm{P}(F_n|B_j)\mathrm{P}(B_j)} = \frac{\left(\frac{i}{k}\right)^n \cdot \frac{1}{k+1}}{\sum_{j=0}^{k} \left(\frac{j}{k}\right)^n \cdot \frac{1}{k+1}} = \frac{\left(\frac{i}{k}\right)^n}{\sum_{j=0}^{k} \left(\frac{j}{k}\right)^n}$$

So the probability that the $(n + 1)$th result is still head is

$$\mathrm{P}(\mathrm{H}|F_n) = \sum_{i=0}^{k} \mathrm{P}(\mathrm{H}|F_n B_i)\mathrm{P}(B_i|F_n) = \frac{\sum_{i=0}^{k}\left(\frac{i}{k}\right)^{n+1}}{\sum_{j=0}^{k}\left(\frac{j}{k}\right)^{n}}$$

But when k is relatively large, the rectangular approximation method of the integral can be used. That is,

$$\frac{1}{k}\sum_{i=0}^{k}\left(\frac{i}{k}\right)^{n+1} \approx \int_0^1 x^{n+1}dx = \frac{1}{n+2}$$

$$\frac{1}{k}\sum_{j=0}^{k}\left(\frac{j}{k}\right)^{n} \approx \int_0^1 x^{n}dx = \frac{1}{n+1}$$

Therefore, when k is relatively large

$$\mathrm{P}(\mathrm{H}|F_n) \approx \frac{n+1}{n+2}$$

Example 12*. (Gambler's bankruptcy problem) Two gamblers, A and B, gamble on the outcome of consecutive tosses of a coin. If head H comes up for each toss, B will pay A one dollar. If tail T comes up, A will pay B one dollar. This continues until one party loses all their money. Let the consecutive tossing of coins be independent, and the probability of heads H appearing each time is p. Let A have i dollar at the beginning, and B have $N - i$ dollar. $N = i + (N - i)$ is the total money of A and B.

(1) Find the probability that A can win money.

(2) Find the probability that B can win money.

Solution. (1) Let E = {A has i dollars and can win money}, denoted as $\mathrm{P(E)} = P_i$. Because {H}, {T} is a partition of the sample space U, we get from the total probability formula

$$P_i = \mathrm{P(E)} = \mathrm{P(E|H)P(H)} + \mathrm{P(E|T)P(T)} = \mathrm{P(E|H)}p + \mathrm{P(E|T)}(1-p)$$

Because under the condition that the head H appears, A wins 1 dollar, that is, A has

$i + 1$ dollars, therefore

$$P(E|H) = P_{i+1}$$

On the contrary, T appears

$$P(E|T) = P_{i-1}$$

Thus we get

$$P_i = pP_{i+1} + (1-p)P_{i-1} \qquad i = 1, 2, \ldots, N-1 \tag{1}$$

The boundary conditions are:

P_0 = P{A has 0 dollars and can win money}= 0

P_N = P{A has N dollars and already won all the money}

= P{Gambling is over} = 1 [definite event]

Also, since $p + q = 1$, equation (1) is equivalent to

$$pP_i + qP_i = pP_{i+1} + qP_{i-1}$$

Or

$$P_{i+1} - P_i = \frac{q}{p}(P_i - P_{i-1}) \qquad i = 1, 2, \ldots, N-1$$

Thus, from $P_0 = 0$ we have

$$P_2 - P_1 = \frac{q}{p}(P_1 - P_0) = \frac{q}{p}P_1$$

$$P_3 - P_2 = \frac{q}{p}(P_2 - P_1) = \left(\frac{q}{p}\right)^2 P_1$$

$$\vdots$$

$$P_i - P_{i-1} = \frac{q}{p}(P_{i-1} - P_{i-2}) = \left(\frac{q}{p}\right)^{i-1} P_1$$

$$\vdots$$

$$P_N - P_{N-1} = \frac{q}{p}(P_{N-1} - P_{N-2}) = \left(\frac{q}{p}\right)^{N-1} P_1$$

Adding up the first $i - 1$ equations, we can get

$$P_i - P_1 = P_1\left[\frac{q}{p} + \left(\frac{q}{p}\right)^2 + \cdots + \left(\frac{q}{p}\right)^{i-1}\right]$$ This is a geometric sequence.

Therefore

$$P_i = \begin{cases} \dfrac{1-\left(\dfrac{q}{p}\right)^i}{1-\dfrac{q}{p}} P_1 & if \quad \dfrac{q}{p} \neq 1 \quad \Rightarrow \quad p \neq \dfrac{1}{2} \\ iP_1 & if \quad \dfrac{q}{p} = 1 \quad \Rightarrow \quad p = \dfrac{1}{2} \end{cases}$$

Using $P_N = 1$, we can get

$$1 = P_N = \begin{cases} \dfrac{1-\left(\dfrac{q}{p}\right)^N}{1-\dfrac{q}{p}} P_1 \\ NP_1 \end{cases}$$

Solving gives

$$P_1 = \begin{cases} \dfrac{1-\dfrac{q}{p}}{1-\left(\dfrac{q}{p}\right)^N} \\ \dfrac{1}{N} \end{cases}$$

$$\text{Hence} \qquad P_i = \begin{cases} \dfrac{1-\left(\dfrac{q}{p}\right)^i}{1-\dfrac{q}{p}} P_1 \\ \\ iP_1 \end{cases} = \begin{cases} \dfrac{1-\left(\dfrac{q}{p}\right)^i}{1-\left(\dfrac{q}{p}\right)^N} & if \quad p \neq \dfrac{1}{2} \\ \\ \dfrac{i}{N} & if \quad p = \dfrac{1}{2} \end{cases}$$

That is, the probability that A can win money is

$$\mathrm{P(E)} = P_i = \begin{cases} \dfrac{1-\left(\dfrac{q}{p}\right)^i}{1-\left(\dfrac{q}{p}\right)^N} & if \quad p \neq \dfrac{1}{2} \\ \\ \dfrac{i}{N} & if \quad p = \dfrac{1}{2} \end{cases}$$

(2) Let G = {B has $N - i$ dollars and can win money}, denoted as P(G) = Q_i. Then, by symmetry, it is only necessary to replace p and q with each other and replace i with $N - i$, so that we have

$$\mathrm{P(G)} - Q_i = \begin{cases} \dfrac{1-\left(\dfrac{p}{q}\right)^{N-i}}{1-\left(\dfrac{p}{q}\right)^N} & if \quad q \neq \dfrac{1}{2} \\ \\ \dfrac{N-i}{N} & if \quad q = \dfrac{1}{2} \end{cases}$$

In addition

$$P_i + Q_i = \begin{cases} \dfrac{1-\left(\dfrac{q}{p}\right)^i}{1-\left(\dfrac{q}{p}\right)^N} + \dfrac{1-\left(\dfrac{p}{q}\right)^{N-i}}{1-\left(\dfrac{p}{q}\right)^N} \\ \\ \dfrac{i}{N} + \dfrac{N-i}{N} \end{cases} = \begin{cases} 1 \\ 1 \end{cases}$$

Therefore we have

$$P(E) + P(G) = 1$$

Now, give a quantitative illustration of the above conclusion. If A has \$5 at the beginning and B has \$10. That is, $N = \$15$. Then when $p = \frac{1}{2}$, the probability that A wins money is $\frac{1}{3}$. And when $p = 0.6$, the probability of A winning money soars to

$$P(E) = \frac{1-\left(\frac{2}{3}\right)^5}{1-\left(\frac{2}{3}\right)^{15}} \approx 0.870$$

Exercise 2.5

1. In a particular population of men and women, 52% are men and 48% are women. 88% of men are right-handed. 92% of women are right-handed. A person is chosen at random from the population.

a. Find the probability that the person is right-handed.

b. Given that the person is right-handed, what is the probability that this person is a woman?

2. A laboratory blood test is 98% effective in detecting a particular disease when it is, in fact, present. However, the test also yields a "false positive result" for 1% of the healthy persons tested. If 0.3% of the population has the disease, what is the probability that a person has the disease, given that the test result is positive?

3. There are two boxes on the table. The first box contains 3 orange and 2 white ping pong balls. The second box contains 2 orange and 5 white ping pong balls. A box is randomly selected, and 1 ping pong ball is randomly drawn. It is known that the ping pong ball drawn is orange. Is it more likely that it comes from the first box or that it comes from the second box?

4. Urn A contains 11 cards numbered 1 through 11, and urn B contains 7 cards numbered 1 through 7. An urn is chosen randomly, and a card is drawn randomly. If the number on the card is even, find the probability that the card comes from urn A.

5. In a school, 5% of the boys and 1% of the girls are taller than 180 cm. 60% of all students at this school are girls. If a student is chosen randomly and found to be taller than 180 cm, what is the probability that this student is a girl?

6. Binary digits "0" and "1" are transmitted over a communication channel. When received by a message station, the probability that a "0" is mistakenly received as a "1" is 0.02. The probability that a "1" is mistakenly received as a "0" is 0.01. The signals "0" and "1" are transmitted with a frequency of 2 : 1. What is the probability that the real signal is a "0" if the received signal is a "0"?

7. There are three coins in the box. One is a fair coin. One is a double-headed coin. One is a weighted coin with a 75% probability of showing heads. Choose a coin at random from the box and throw it away. Given that it shows a head, what is the probability that it is a double-headed coin?

8. A factory produces the same product in halls No 1, 2, and 3, and the output of each hall accounts for $\frac{1}{2}$, $\frac{1}{3}$, and $\frac{1}{6}$ of the total production, respectively. The defective rates are 1%, 1%, and 2% respectively. Now, select a product at random from the factory's products.

a. What is the probability that the product is defective?

b. Given that the product is defective, what is the probability that it comes from No 1 hall?

9. The probability that a person with hepatitis B will be diagnosed by examination is 0.98. The probability that a normal person will be misdiagnosed as having hepatitis B by examination is 0.05. 5% of the population in a city has hepatitis B. One person is randomly selected from the residents of the city for examination. If the person is diagnosed as having hepatitis B, find the probability that the person actually has hepatitis B.

10. There are three pen holders on the table. The first one contains 2 red ballpoint pens and 4 blue ballpoint pens. The second one contains 4 red ballpoint pens and 2 blue ballpoint pens. The third one contains 3 red ballpoint pens and 3 blue ballpoint pens. They look the same. Choose one pen holder at random first, then take one ballpoint pen at random.

a. Find the probability that the ballpoint pen taken is red.

b. Given that the ballpoint pen taken is red, which pen holder does the pen have the greatest probability of coming from?

11. In a town, 30% of the people are conservatives, 50% socialists and 20% liberals. In this town, at the last election, 65% of the conservatives voted, as did 82% of the socialists and 50% of the liberals. A person from the town is selected randomly and votes in the last election. What is the probability that this person is a liberal?

12. The warehouse stock consists of boxes of excellent, good and low-quality towels in proportion to 1 : 3 : 2, respectively. The probabilities of unsatisfactory towels for these three types are 0, 0.1, and 0.3, respectively. A box is chosen randomly, and two towels are checked and found to be satisfactory. What is the probability that it is

- **a.** of good quality?
- **b.** of low quality?

13. Three boys, Antti, Ben and Charles, wash the family dishes. Antti is the oldest. He does the job 40% of the time. Ben and Charles share the other 60% equally. The probability that at least one dish will be broken when Antti is washing them is 0.01; for Ben and Charles, the probabilities are 0.03 and 0.02. The parents do not know who is washing the dishes, and one night, they heard one dish break. What is the probability that Antti was washing it?

14. Urn A contains 8 black balls and 4 white balls. Urn B contains 5 black balls and 6 white balls. Two balls are taken from A to B, and then one ball is drawn from B.

- **a.** What is the probability that the ball drawn is white?
- **b.** Given that the ball drawn is white, what is the probability that at least one white ball was taken from urn A?

15. An automobile insurance company classifies drivers as class A (good risks), class B (medium risks) and class C (poor risks). They believe that class A risks constitute 30% of the drivers who buy car insurance from them, class B 50% and class C 20%. The probability that a class A driver will have one or more accidents in any 12 months is 0.01. For a class B driver, the probability is 0.03, and for a class C driver, it is 0.1. A new driver just bought a new car insurance, and within 12 months, the driver had an accident. What is the probability that the driver is a class B driver?

Chapter 3 Random variables and their distributions

3.1. Random variables, Probability distributions of discrete random variables

Random variables

To study the outcomes of random experiments comprehensively and reveal the objective statistical laws, we connect the outcomes of random experiments with real numbers and numericalise the outcomes of random experiments. Hence, we introduce the concept of random variables.

Consider the random experiment of "tossing a coin". It has two possible outcomes – "getting a head" or "getting a tail". To facilitate research, we use a real number to represent an outcome. For example, use the number "1" to represent "getting a head" and the number "0" to represent "getting a tail". When we discuss the experiment's outcomes, we can simply say that the outcome is either number 1 or number 0. After this relationship is established, in fact a real-value random variable X is introduced. For the two outcomes of the experiment, we set the values of X to 1 and 0, respectively. The random variable X takes different values when the experiment gives different outcomes. If we connect it to the sample space of an experiment, then corresponding to different outcomes of the sample space, the random variable X takes different values. Therefore, X is a function defined on the sample space U, i.e.

Sample space $U = \{e\} = \{\text{head}, \ \text{tail}\}$

numericalise $X(e) = X$ ($X: e \to X$) ↓ ↓ ↓ ↓

numericalisation of sample space Range $R_X = \{X\} = \{1, \ 0\}$

So $e = \begin{cases} head \\ tail \end{cases} \xleftrightarrow{\ \ numericalisation\ \ } \mathrm{X}(e) = \mathrm{X} = \begin{cases} 1 & (1 \ \ represents \ \ head.) \\ 0 & (0 \ \ represents \ \ tail.) \end{cases}$

After introducing the random variable X, we can use random variables to describe events. When X takes the value 1, denote it as $\{X = 1\}$. It represents the event $\{\text{head}\}$. When X takes the value 0, denote it as $\{X = 0\}$. It represents the event $\{\text{tail}\}$. Therefore

$$P\{X = 1\} = P\{H\} = 0.5$$
$$P\{X = 0\} = P\{T\} = 0.5$$

Definition

Let the sample space U of a random experiment S be $U = \{e\}$. If there is a real number X(e) corresponding to each $e \in U$, then we get a real-value function defined on the sample space U, X(e) = X. This is the so-called random variable.

By definition, we have

sample space $\mathrm{U} = \{\ e\ \} \ = \ \{e_1, e_2, \dots, e_n, \dots\}$

numericalise X(e) = X (X: $e \to$ X)

Range $\mathrm{R_X} = \{\mathrm{X}\} \ = \ \{x_1, x_2, \dots, x_n, \dots\}$

So $\quad e = e_1, e_2, \dots, e_n, \dots \xleftrightarrow{\ \ numericalisation\ \ } \mathrm{X} = x_1, x_2, \dots, x_n, \dots$

Or it can also be expressed as

$$e = \begin{cases} e_1 \\ e_2 \\ \vdots \\ e_n \\ \vdots \end{cases} \xleftrightarrow{\ \ numericalisation\ \ } \mathrm{X} = \begin{cases} x_1 & (x_1 \ \ represents \ \ e_1.) \\ x_2 & (x_2 \ \ represents \ \ e_2.) \\ \vdots & \vdots \\ x_n & (x_n \ \ represents \ \ e_n.) \\ \vdots & \vdots \end{cases}$$

The above two representation methods are equivalent.

The outcomes of many experiments are themselves real numbers. That is, the outcome itself is the real number X. At this time, we naturally define $X = x$.

For example, consider the experiment "rolling a die". The sample space $U = \{1, 2, \ldots, 6\}$.

So $\quad R_x = U = \{1, 2, \ldots, 6\} \quad X = \begin{cases} 1 \\ 2 \\ \vdots \\ 6 \end{cases}$

Probability distributions of discrete random variables

For some random variables, the total number of values it can take is finite or countably infinite. Such random variables are called discrete random variables. For instance, the random variable in the previous section can only take two values, 1 and 0, so it is a discrete random variable.

It is easy to know that if we want to find out the statistical law of a discrete random variable, we must know all the possible values that the random variable X can take and the probability of taking each value.

Let all the possible values that the discrete random variable X can take be x_k $(k = 1, 2, \ldots)$.

That is, $\quad X = \begin{cases} x_1 \\ x_2 = x_k \\ \vdots \end{cases} \quad (k = 1, 2, \ldots)$

The probability that X takes each possible value, that is, the occurrence probability of the event $\{X = x_k\}$, is

$$P\{X = x_k\} = p_k \qquad k = 1, 2, \ldots$$

This is called the probability distribution or distribution of the discrete random variable X.

The probability distribution can be represented in a table.

X	x_1	x_2	$\cdots$	x_n	$\cdots$
p_k	p_1	p_2	$\cdots$	p_n	$\cdots$

According to the definition of probability, p_k satisfies the following two conditions:

1. $p_k \geq 0, \quad k = 1, 2, \ldots$

2. $\mathrm{P}\{X \geq x_1\} = \sum_{k=1}^{\infty} \mathrm{P}\{X = x_k\} = \sum_{k=1}^{\infty} p_k = 1 \qquad k = 1, 2, \ldots$

Example. Tom goes to work by car. On his way to work, he needs to pass four traffic lights. The probability of letting a car pass or stop at each traffic light is $\frac{1}{2}$. Let X be the number of traffic lights the car passes when it stops for the first time (all traffic lights work independently). Find the distribution of the random variable X.

Solution. Let $p = \frac{1}{2}$ be the probability of letting a car pass at each traffic light,

$(1 - p) = \frac{1}{2}$ be the probability of stopping a car at each traffic light.

In the following, let us analyse.

(1) The event $\{X = 0\}$ means "the car stops at the first traffic light". So

$$\mathrm{P}\{X = 0\} = 1 - p = \frac{1}{2} = 0.5$$

(2) The event $\{X = 1\}$ means "the car passes the first traffic light and stops at the second traffic light". (These are product events that occur

simultaneously.) So

$$P\{X=1\} = p(1-p) = \frac{1}{2} \times \frac{1}{2} = \frac{1}{4} = 0.25$$

(3) The event $\{X=2\}$ means "the car passes the first two traffic lights and stops at the third traffic light". (These are product events that occur simultaneously.) So

$$P\{X=2\} = pp(1-p) = \frac{1}{2} \times \frac{1}{2} \times \frac{1}{2} = \frac{1}{8} = 0.125$$

(4) The event $\{X=3\}$ means "the car passes the first three traffic lights and stops at the fourth traffic light". (These are product events that occur simultaneously.) So

$$P\{X=3\} = ppp(1-p) = \frac{1}{2} \times \frac{1}{2} \times \frac{1}{2} \times \frac{1}{2} = \frac{1}{16} = 0.0625$$

(5) The event $\{X=4\}$ means "the car passes all four traffic lights". (These are product events that occur simultaneously.) So

$$P\{X=4\} = pppp = \frac{1}{2} \times \frac{1}{2} \times \frac{1}{2} \times \frac{1}{2} = \frac{1}{16} = 0.0625$$

Thus, we get the distribution of X.

$$\begin{cases} P\{X=k\} = p^k(1-p) & k=0,1,2,3 \\ P\{X=4\} = p^4 & \end{cases}$$

The distribution can be represented in a table.

X	0	1	2	3	4
p_k	$1-p$	$p(1-p)$	$pp(1-p)$	$ppp(1-p)$	p^4

Substituting $p=\frac{1}{2}$ we get

X	0	1	2	3	4
p_k	0.5	0.25	0.125	0.0625	0.0625

Exercise 3.1

1. A discrete random variable X has the following probability distribution.

X	0	2	5	7
p_k	$\frac{1}{6}$	$\frac{1}{3}$	$\frac{1}{4}$	a

Find the value of a.

2. A discrete random variable Y has the following probability distribution.

Y	1	3	5	7	9
p_k	2k	3k	4k	5k	6k

a. Find the value of k.

b. Find $P\{Y = 3\}$.

c. Find $P\{Y > 3\}$.

3. There are 5 balls in a bag, numbered 1, 2, 3, 4, 5. Take 3 balls randomly from the bag. Let X represent the largest number among the 3 balls drawn. Find the probability distribution of X.

4. Let the probability distribution of a random variable X be $P\{X = k\} = \dfrac{a}{k+2}$, $k = 0, 1, 2, 3$. Find the value of a.

5. Among 20 items of the same type, 2 were defective. Randomly select 3 items, one after another, without replacement. Let X represent the number of defective items taken.

a. Find the probability distribution of X.

b. Display the distribution graphically.

6. Let the probability distribution of a random variable X be

$$P\{X = x\} = k\left(\frac{5}{6}\right)^x \qquad x = 1, 2, \ldots$$

Find the value of *k*.

7. Five devices are subject to successive reliability tests. Each device is tested only if the preceding one turns out to be reliable. The probability of passing the test is 0.9 for each device. Let X represent the number of tests done. Find the probability distribution of X.

3.2. Four important probability distributions of discrete random variables

3.2.1. (0 – 1) distribution

If the random variable X can only take two values, 0 and 1, that is,

$$X = k = \begin{cases} 0 & \quad 1-p \\ 1 & \quad p \end{cases}$$

its probability distribution is

$$P\{X = k\} = p^k(1-p)^{1-k}, \qquad k = 0, 1 \quad (0 < p < 1)$$

then we say that X has a (0 – 1) distribution.

The (0 – 1) distribution can be also illustrated as

X	0	1
p_k	$1-p$	p

For example, the experiment "tossing a coin" can be described by a random variable of (0 – 1) distribution. That is,

$$X = k = \begin{cases} 0 & (0 \quad represents \quad tail.) \\ 1 & (1 \quad represents \quad head.) \end{cases}$$

The (0 – 1) distribution is an important distribution that we often encounter.

3.2.2. Binomial distribution

Repeat n times the experiment S. If the outcomes of each experiment do not affect each other, that is, the probability of the outcomes of each experiment does not depend on the outcomes of other experiments, then we will say these n times experiments are mutually independent.

If the experiment S has only two possible outcomes: A and A', P(A) = p, P(A') = $1 - p = q$ ($0 < p < 1, 0 < q < 1, p + q = 1$), repeat n times the experiment S, then we call this series of repeated independent experiments as n times Bernoulli experiment, or simply Bernoulli experiment for short. It is an experiment of sampling with replacement. Bernoulli experiment is a very important mathematical model with a wide range of applications. It is one of the most studied models. In the following, we use an example of this mathematical model to derive its probability formula: the probability formula for the binomial distribution.

Example. Toss a coin $n = 3$ times. Find the probability that the head occurs $k = 2$ times.

Solution. Let the outcome of each coin toss be X_i ($i = 1, 2, 3$). Then

$$X_i = \begin{cases} 0 & \{T\} \quad occurs & 1-p=\dfrac{1}{2} \\ 1 & \{H\} \quad occurs & p=\dfrac{1}{2} \end{cases}$$ follows a (0 – 1) distribution.

Let X represent the number of times the event {H} occurs in tossing a coin three times. Then

$$X = \sum_{i=1}^{3} X_i = k = \begin{cases} 0 \\ 1 \\ 2 \\ 3 \end{cases}$$

Since event {X = 2} occurs = $\begin{cases} \text{HHT} & occurs. \\ & or \\ \text{HTH} & occurs. \\ & or \\ \text{THH} & occurs. \end{cases}$

so, tossing a coin 3 times, the probability of getting heads {H} twice is

$$\begin{aligned} P\{X = 2\} &= pp(1-p) + p(1-p)p + (1-p)pp \\ &= 3p^2(1-p) \\ &= \binom{3}{2} p^2(1-p)^{3-2} \end{aligned}$$

where summation terms $\binom{3}{2} = 3.$

Therefore $P\{X = 2\} = \binom{3}{2} p^2(1-p)^{3-2}$ □

Deriving from the above example and by induction, we can get the formula.

$$P\{X = k\} = \binom{n}{k} p^k(1 - p)^{n - k} \qquad X = \sum_{i=1}^{3} X_i = k = \begin{cases} 0 \\ 1 \\ 2 \\ \vdots \\ n \end{cases}$$

This formula is called the probability formula for the binomial distribution.

Obviously $\quad P\{X - k\} \geq 0$

$$P\{0 \leq X \leq n\} = \sum_{k=0}^{n} P\{X = k\} = \sum_{k=0}^{n} \binom{n}{k} p^k (1 - p)^{n-k} = (p + q)^n = 1^n = 1$$

Since the binomial theorem formula is

$$(p + q)^n = \sum_{k=0}^{n} \binom{n}{k} p^k (1 - p)^{n-k}$$

In the expansion, the term with p^k is $\binom{n}{k} p^k (1 - p)^{n-k}$, so we say that the random variable X has a binomial distribution with parameters n and p, denoted as $X \sim B(n, p)$.

Especially when $n = 1$, the binomial distribution becomes the (0 – 1) distribution.

$$P\{X = k\} = \binom{1}{k} p^k q^{1 - k} = p^k q^{1 - k} \qquad X = k = \begin{cases} 0 \\ 1 \end{cases}$$

Based on the above, we can give the following definition:

(1) The method of the experiment S is sampling with replacement or equivalent to sampling with replacement (e.g., tossing a coin). In this way, the outcomes of n repeated experiments $X_1, X_2, \ldots, X_n$ are mutually independent.

(2) There are only two outcomes for each experiment S.

$$X_i = \begin{cases} 0 & 1-p \\ 1 & p \end{cases} \qquad \text{follows a } (0-1) \text{ distribution.}$$

And each X_i ($i = 1, 2,\ldots, n$) has the same probability distribution.

Then we refer to this series of experiments as n times Bernoulli experiments, or simply Bernoulli experiments for short. It is an experiment of sampling with replacement. In n times Bernoulli experiments, the number X of occurrences of event $A = \{X_i = 1\}$ is called a random variable of a binomial distribution, denoted as $X \sim B(n, p)$. **The probability formula of a binomial distribution is**

$$P\{X = k\} = \binom{n}{k} p^k (1-p)^{n-k} \qquad X = \sum_{i=1}^{3} X_i = k = \begin{cases} 0 \\ 1 \\ 2 \\ \vdots \\ n \end{cases}$$

Moreover $$1 = P\{0 \le X \le n\} = \sum_{k=0}^{n} P\{X = k\} = \sum_{k=0}^{n} \binom{n}{k} p^k (1-p)^{n-k}$$ □

Example 1. Roll a die five times. Find the probability of observing three "2s".

Solution. Since the outcome of each throw

$$X_i = \begin{cases} 0 & (0 \;\; \textit{represents} \;\; 1,3,4,5,6.) \\ 1 & (1 \;\; \textit{represents} \;\; 2.) \end{cases} \qquad (i = 1, 2, 3, 4, 5)$$

so $P(A) = P\{X_i = 1\} = p = \dfrac{1}{6}$ $\qquad P(A') = P\{X_i = 0\} = 1 - p = 1 - \dfrac{1}{6}$

In five throws, the number of times "2" appears is $X = \sum_{i=1}^{n} X_i = \sum_{i=1}^{5} X_i = k = \begin{cases} 0 \\ 1 \\ 2 \\ 3 \\ 4 \\ 5 \end{cases}$

That is, $X = k = \begin{cases} 0 \\ 1 \\ 2 \\ 3 \\ 4 \\ 5 \end{cases} \sim B(5, \frac{1}{6})$ follows a binomial distribution.

The formula is $P\{X = k\} = \binom{n}{k} p^k (1-p)^{n-k}$.

Therefore, the probability of the event $\{X = 3\}$ occurring is

$$P\{X = 3\} = \binom{5}{3} (\frac{1}{6})^3 (1 - \frac{1}{6})^{5-3} \approx 0.0322$$

Example 2. The probability of a successful free throw in basketball is 80%. Find the probability that at least 3 out of 5 free throws are successful.

Solution. The outcome of each throw

$$X_i = \begin{cases} 0 & (0 \quad represents \quad 'failure') \\ 1 & (1 \quad represents \quad 'success'.) \end{cases}$$

$p = 80\% = 0.8 \qquad q = 1 - p = 1 - 0.8 = 0.2$

Number of "successful" free throws out of five $X = \sum_{i=1}^{n} X_i = \sum_{i=1}^{5} X_i = k = \begin{cases} 0 \\ 1 \\ 2 \\ 3 \\ 4 \\ 5 \end{cases}$

Because $$P\{m \le X \le n\} = \sum_{k=m}^{n} P\{X = k\}$$

So the probability of the event {at least three successful free throws out of five} = $\{3 \le X \le 5\}$ is

$$P\{3 \le X \le 5\} = \sum_{k=3}^{5} P\{X = k\} = P\{X = 3\} + P\{X = 4\} + P\{X = 5\}$$

That is, $$P\{3 \le X \le 5\} = \binom{5}{3} 0.8^3 0.2^2 + \binom{5}{4} 0.8^4 0.2^1 + \binom{5}{5} 0.8^5 0.2^0 \approx 0.942$$

Example 3. Jarkko has shooting practice. The probability of each shot hitting the target is 0.02. He fires 400 times independently. Find the probability of hitting the target at least twice.

Solution. Every shot can be considered as an experiment. If the number of times the target is hit is X, then X ~ B (400, 0.02). The probability distribution of X is

$$P\{X = k\} = \binom{n}{k} p^k (1-p)^{n-k} = \binom{400}{k} 0.02^k (1-0.02)^{400-k} \qquad X = k = \begin{cases} 0 \\ 1 \\ 2 \\ \vdots \\ 400 \end{cases}$$

Since $$1 = P\{0 \le X \le 400\} = \sum_{k=0}^{400} P\{X = k\} = \sum_{k=0}^{1} P\{X = k\} + \sum_{k=2}^{400} P\{X = k\}$$

$$= P\{0 \le X \le 1\} + P\{2 \le X \le 400\}$$

so $$\begin{aligned} P\{X \ge 2\} &= 1 - P\{0 \le X \le 1\} \\ &= 1 - P\{X = 0\} - P\{X = 1\} \\ &= 1 - \binom{400}{0} 0.02^0 0.98^{400-0} - \binom{400}{1} 0.02^1 0.98^{400-1} \\ &= 1 - 0.98^{400} - 400 \cdot 0.02 \cdot 0.98^{399} \approx 0.997 \end{aligned}$$

Example 4. A language listening test consists of multiple-choice questions. Each question has four choices, only one of which is correct. How many questions must be answered so that the probability of at least two being correct exceeds 0.9?

Solution. For each question

$$X_i = \begin{cases} 0 \quad (0 \text{ represents the three incorrect answers.}) & q = 1 - p = \dfrac{3}{4} \\ 1 \quad (1 \text{ represents the only correct answer.}) & p = \dfrac{1}{4} \end{cases}$$

In n answers, the number of correct ones $\qquad X = \sum_{i=1}^{n} X_i = k = \begin{cases} 0 \\ 1 \\ 2 \\ \vdots \\ n \end{cases}$

The probability of the event {at least two answers are correct} = $\{2 \le X \le n\}$

occurring is $\qquad P\{2 \le X \le n\} = \sum_{k=2}^{n} P\{X = k\}$.

Because $\qquad 1 = P\{0 \le X \le n\} = \sum_{k=0}^{n} P\{X = k\} = \sum_{k=0}^{1} P\{X = k\} + \sum_{k=2}^{n} P\{X = k\}$

so $\qquad \sum_{k=2}^{n} P\{X = k\} = 1 - \sum_{k=0}^{1} P\{X = k\}$

That is, $\qquad P\{2 \le X \le n\} = 1 - \sum_{k=0}^{1} P\{X = k\} = 1 - [\binom{n}{0} p^0 q^{n-0} + \binom{n}{1} p^1 q^{n-1}]$

$$= 1 - [\binom{n}{0}\left(\frac{1}{4}\right)^0\left(\frac{3}{4}\right)^{n-0} + \binom{n}{1}\left(\frac{1}{4}\right)^1\left(\frac{3}{4}\right)^{n-1}] = 1 - \frac{3^{n-1}}{4^n}(3 + n)$$

Since it must satisfy $\qquad P\{2 \le X \le n\} > 0.9$

so $\qquad 1 - \dfrac{3^{n-1}}{4^n}(3 + n) > 0.9$

That is, $\qquad \dfrac{3^{n-1}}{4^n}(3 + n) < 0.1$

Solve by trial, when $n = 14$, left side ≈ 0.10097; when $n = 15$, left side ≈ 0.0802. So, at least 15 questions must be answered. (Or solve graphically— draw the graph of $y = 0.1 - \frac{3^{n-1}}{4^n}(3 + n)$, then find zero.)

3.2.3. Poisson distribution

Let us read one example first.

Example. A person has shooting practice. The probability of each shot hitting the target is 0.02. He fired 400 times independently. Find the probability of hitting the target at least twice.

Solution. Every shot can be considered as an experiment. If the number of times the target is hit is X, then X ~ B (400, 0.02).

The probability distribution of X is

$$P\{X = k\} = \binom{n}{k} p^k(1-p)^{n-k} = \binom{400}{k} 0.02^k(1-0.02)^{400-k}, \quad X = k = \begin{cases} 0 \\ 1 \\ 2 \\ \vdots \\ 400 \end{cases}$$

So
$$\begin{aligned} P\{X \geq 2\} &= 1 - P\{0 \leq X \leq 1\} \\ &= 1 - P\{X = 0\} - P\{X = 1\} \\ &= 1 - \binom{400}{0} 0.02^0(1-0.02)^{400-0} - \binom{400}{1} 0.02^1(1-0.02)^{400-1} \\ &= 1 - 0.98^{400} - 400(0.02)(0.98)^{399} \end{aligned}$$

It is complicated to calculate the above expression directly without a calculator.

Below, we give an approximate calculation formula when n is large and p is small. This is the famous Poisson approximation of the binomial distribution.

Poisson theorem

If $np = \lambda > 0$ is a constant, n is a positive integer, then to any given non-negative integer k, we have

$$\lim_{n\to\infty}\binom{n}{k}p^k(1-p)^{n-k} = \frac{\lambda^k e^{-\lambda}}{k!}$$

Proof. From $p = \dfrac{\lambda}{n}$, we get

$$\binom{n}{k}p^k(1-p)^{n-k} = \frac{n(n-1)\cdots(n-k+1)}{k!}\left(\frac{\lambda}{n}\right)^k\left(1-\frac{\lambda}{n}\right)^{n-k}$$

$$= \frac{\lambda^k}{k!}[1\left(1-\frac{1}{n}\right)\left(1-\frac{2}{n}\right)\cdots\left(1-\frac{k-1}{n}\right)]\left(1-\frac{\lambda}{n}\right)^n\left(1-\frac{\lambda}{n}\right)^{-k}$$

For any given value of k, when $n \to \infty$

$$[1\left(1-\frac{1}{n}\right)\left(1-\frac{2}{n}\right)\cdots\left(1-\frac{k-1}{n}\right)] \to 1$$

$$\left(1-\frac{\lambda}{n}\right)^n \to e^{-\lambda} \qquad \left(1-\frac{\lambda}{n}\right)^{-k} \to 1$$

Therefore, we have

$$\lim_{n\to\infty}\binom{n}{k}p^k(1-p)^{n-k} = \frac{\lambda^k e^{-\lambda}}{k!}$$

Obviously, the condition of the theorem $np = \lambda$ (constant) means that when n is large, p must be small. Therefore, the theorem above shows that we have the following approximation when n is very large and p is very small.

$$\binom{n}{k} p^k(1-p)^{n-k} \approx \frac{\lambda^k e^{-\lambda}}{k!} \qquad \text{where} \qquad \lambda = np$$

In practical calculations, when $n \geq 20, p \leq 0.05$, $\dfrac{\lambda^k e^{-\lambda}}{k!}$ $(\lambda = np)$ is a good approximation to $\binom{n}{k} p^k(1-p)^{n-k}$. When $n \geq 100$, $np \leq 10$, the approximation effect is better.

We use this approximate formula to calculate the probability of $P\{X \geq 2\}$ in the above example.

$$P\{X = k\} = \binom{n}{k} p^k(1-p)^{n-k} \approx \frac{\lambda^k e^{-\lambda}}{k!}, \qquad \lambda = np = 400 \cdot 0.02 = 8$$

So $\qquad P\{X = 0\} \approx \dfrac{8^0 e^{-8}}{0!} = e^{-8}, \qquad P\{X = 1\} \approx \dfrac{8^1 e^{-8}}{1!} = 8e^{-8}$

Therefore $\qquad P\{X \geq 2\} \approx 1 - e^{-8} - 8e^{-8} \approx 0.997$

Definition of Poisson distribution

If all the possible values that a random variable X can take are 0, 1, 2,…, and the probability of taking each value is

$$P\{X = k\} = \frac{\lambda^k e^{-\lambda}}{k!}, \qquad X = k = \begin{cases} 0 \\ 1 \\ 2 \\ \vdots \end{cases}$$

where $\lambda = np > 0$ is a constant.

then we say X has a Poisson distribution with parameter λ, denoted as $X \sim Po(\lambda)$.

It is easy to know

$$P\{X = k\} \geq 0, \qquad X = k = \begin{cases} 0 \\ 1 \\ 2 \\ \vdots \end{cases}$$

And $$P\{X \geq 0\} = \sum_{k=0}^{\infty} P\{X = k\} = \sum_{k=0}^{\infty} \frac{\lambda^k e^{-\lambda}}{k!} = e^{-\lambda} \sum_{k=0}^{\infty} \frac{\lambda^k}{k!} = e^{-\lambda} \cdot e^{\lambda} = 1$$

Example 1. Suppose the number of typographical errors on a book page follows a Poisson distribution with parameter $\lambda = \frac{2}{5}$. Find the probability that there are at least two errors on this page.

Solution. Let X represent the number of errors on this page, then the probability that there are at least two errors is

$$P\{X \geq 2\} = 1 - P\{X = 0\} - P\{X = 1\} = 1 - \frac{\left(\frac{2}{5}\right)^0 e^{-\frac{2}{5}}}{0!} - \frac{\left(\frac{2}{5}\right)^1 e^{-\frac{2}{5}}}{1!} \approx 0.0616$$

That is, $P\{X \geq 2\} \approx 0.0616$.

Example 2. If the probability of defective components produced by a particular machine is 0.05, find the probability that there are at most 3 defective components out of 100. Calculate using Poisson's theorem.

Solution. Let each component be

$$X_i = \begin{cases} 1 & \textit{defective} & p = 0.05 \\ 0 & \textit{non-defective} & 1 - p = 0.95 \end{cases} \qquad i = 1, 2, \ldots, 100$$

Because X_i follows a $(0 - 1)$ distribution, the number of defectives is

$$X = \sum_{i=1}^{100} X_i \sim B(100, 0.05) \quad \text{follows a binomial distribution.}$$

If we calculate with Poisson's theorem, then the parameter is

$$\lambda = np = 100 \cdot 0.05 = 5$$

Therefore, the probability of at most 3 defectives is

$$P\{0 \le X \le 3\} = \frac{5^0 e^{-5}}{0!} + \frac{5^1 e^{-5}}{1!} + \frac{5^2 e^{-5}}{2!} + \frac{5^3 e^{-5}}{3!} \approx 0.265$$

Example 3. Find the value of k such that the Poisson distribution $P\{X = k\} = \dfrac{\lambda^k e^{-\lambda}}{k!}$ has the maximum value.

Solution. For the relationship between the maximum $P_{max}\{X = k\}$ and the probabilities of two adjacent points, we definitely have

$$P_{max}\{X = k\} \ge P\{X = k - 1\}$$

$$P_{max}\{X = k\} \ge P\{X = k + 1\}$$

So $\quad \dfrac{P_{max}\{X = k\}}{P\{X = k-1\}} \ge 1 \qquad \dfrac{P_{max}\{X = k\}}{P\{X = k+1\}} \ge 1$

Since $\quad \dfrac{P_{max}\{X = k\}}{P\{X = k-1\}} = \dfrac{\dfrac{\lambda^k e^{-\lambda}}{k!}}{\dfrac{\lambda^{k-1} e^{-\lambda}}{(k-1)!}} = \dfrac{\lambda}{k} \qquad \dfrac{P_{max}\{X = k\}}{P\{X = k+1\}} = \dfrac{\dfrac{\lambda^k e^{-\lambda}}{k!}}{\dfrac{\lambda^{k+1} e^{-\lambda}}{(k+1)!}} = \dfrac{k+1}{\lambda}$

so $\quad \dfrac{\lambda}{k} \ge 1 \qquad \dfrac{k+1}{\lambda} \ge 1$

That is, $\quad \lambda \ge k \qquad k \ge \lambda - 1$

Therefore $\quad \lambda - 1 \le k \le \lambda$

Since k is an integer, $\lambda - 1$ and λ are two adjacent integers, therefore, when $k = \lambda$ or $k = \lambda - 1$, the Poisson distribution has maximum values.

$$P_{max} = P\{X = \lambda\} = P\{X = \lambda - 1\}$$

When λ is not an integer, change it to an integer according to the situation.

3.2.4. Hypergeometric distribution

Let us read one example first.

Example. There are 20 products in a box, 4 of which are defective and 16 are non-defective.

(a) Take one product at a time randomly without replacement and repeat 10 times. Find the probability that exactly 8 of the 10 selected products are non-defective.

(b) Take ten products at a time randomly. Find the probability that exactly 8 of the 10 selected products are non-defective.

Solution. (a) Take one product randomly at one time, and the outcome of the product will be X_i.

$$X_i = \begin{cases} 0 & (0 \quad \textit{represents} \quad \textit{defective} \quad \textit{product.}) \\ 1 & (1 \quad \textit{represents} \quad \textit{non-defective} \quad \textit{product.}) \end{cases}$$

Take 10 products without replacement, of which the number of good products is X.

$$X = \sum_{i=1}^{10} X_i = k = \begin{cases} 0 \\ 1 \\ 2 \\ \vdots \\ 10 \end{cases}$$

Since it is sampling without replacement, the outcomes of each random sampling X_i are not independent of the outcomes of other samplings. They have different probability distributions. That is, $X_1, X_2, \ldots$, and X_n are not mutually independent and have different probability distributions.

So $$P\{X=8\} = \binom{10}{8}\frac{\binom{16}{1}\binom{15}{1}\binom{14}{1}\binom{13}{1}\binom{12}{1}\binom{11}{1}\binom{10}{1}\binom{9}{1}\binom{4}{1}\binom{3}{1}}{\binom{20}{1}\binom{19}{1}\binom{18}{1}\binom{17}{1}\binom{16}{1}\binom{15}{1}\binom{14}{1}\binom{13}{1}\binom{12}{1}\binom{11}{1}}$$

$$= \frac{\binom{16}{8}\binom{4}{2}}{\binom{20}{10}} \approx 0.418 \qquad (1)$$

Sampling without replacement in the hypergeometric distribution differs from the sampling with replacement in the binomial distribution. In the binomial distribution, the outcomes of each sampling X_i and the outcomes of other samplings are independent and have the same probability distribution (because it is sampling with replacement). However, their number of possibilities is the same, so the summation terms $\binom{10}{8}$ above are determined according to this property.

(b) Take 10 products at a time, among which the number of good products is

$$Y = k = \begin{cases} 0 \\ 1 \\ 2 \\ \vdots \\ 10 \end{cases}$$

$$P\{Y=8\} = \frac{\binom{16}{8}\binom{4}{2}}{\binom{20}{10}} \approx 0.418$$

It can be seen from (a) and (b) that the probability $P\{X=8\}$, for taking one product at one time without replacement 10 times, of which there are exactly 8 non-defectives and the probability $P\{Y = 8\}$, for taking 10 products at one time, of which there are exactly 8 non-defectives, are the same. That is,

$$P\{X = 8\} = P\{Y = 8\} = \frac{\binom{16}{8}\binom{4}{2}}{\binom{20}{10}} \approx 0.418$$

From this example, we can get the following definition:

If, in each sampling, one object is taken and there are only two possible outcomes.

$$X_i = \begin{cases} 0 & (0 \quad represents \quad e_1.) \\ 1 & (1 \quad represents \quad e_2.) \end{cases}$$

Repeat n times without replacement, and select n objects, among which there are precisely $\{X = k\}$. The probability of $\{X = k\}$ is the same as the probability of selecting n objects simultaneously and having exactly $\{Y = k\}$ among them.

That is, $\qquad P\{X = k\} = P\{Y = k\}$

In this case, we do not need to differentiate between two sampling methods. We can use the same probability formula.

$$P\{X = k\} = \frac{\binom{M}{k}\binom{N-M}{n-k}}{\binom{N}{n}} \qquad X = k = \begin{cases} 0 \\ 1 \\ 2 \\ \vdots \\ n \end{cases} \qquad (n \le M)$$

In the formula, N, M, (N – M) means that the total number of all objects is N pieces, among which M pieces of e_1 and (N – M) pieces of e_2, $N = M + (N - M)$.

n, k, $(n - k)$ means that n pieces are randomly selected from N pieces, and there are exactly k pieces e_1 and $(n - k)$ pieces e_2. $n = k + (n - k)$.

This formula is the **probability formula of the hypergeometric distribution**.

In particular, equation (1) in the above example shows that when M and N are both very large relative to k and n, sampling without replacement can be approximately treated as sampling with replacement. At this time, the hypergeometric distribution of X is approximately equal to the binomial distribution.

That is,
$$P\{X = k\} = \frac{\binom{M}{k}\binom{N-M}{n-k}}{\binom{N}{n}} \approx \binom{n}{k} p^k (1-p)^{n-k}$$

where $p = \dfrac{M}{N}$. M and N are both very large relative to k and n.

Example 1. In the Viking Lottery, 6 numbers are drawn from the numbers 1 to 48. Additionally, 3 additional numbers are drawn. Find the probability

(a) a lottery ticket has 4 numbers that are the same as 4 of the 6 winning numbers;

(b) a lottery ticket has 5 numbers that are the same as 5 of the 6 winning numbers and 1 number that is the same as 1 of the additional 3 winning numbers (5 + 1).

Solution. In the Viking Lottery draw, there are two experiment processes.

(1) In the first process of the experiment, take a number without replacement. And take 6 times. Therefore, the outcome of each random choice X_i is not independent of the outcomes of other choices, and they have different probability distributions.

The outcome of each selection (that is, one number is selected) is X_i.

$$X_i = \begin{cases} 0 & (0 \ \ represents \ \ non-drawn \ \ numbers.) \\ 1 & (1 \ \ represents \ \ drawn \ \ numbers.) \end{cases}$$

Take a number without replacement and take 6 times. The outcome is X.

$$X = \sum_{i=1}^{6} X_i = k = \begin{cases} 0 \\ 1 \\ 2 \\ \vdots \\ 6 \end{cases}$$

(2) In the second process of the experiment, take a number without replacement and take 3 times. Therefore, the result of each random selection Y_i is not independent of the results of other selections, and they have different probability distributions.

The outcome of each selection (that is, one additional number is selected) is $Y_{i.}$

$$Y_i = \begin{cases} 0 & (0 \ \ represents \ \ non-drawn \ \ additional \ \ numbers.) \\ 1 & (1 \ \ represents \ \ drawn \ \ additional \ \ numbers.) \end{cases}$$

Take a number without replacement 3 times. The outcome is Y.

$$Y = \sum_{i=1}^{3} Y_i = k = \begin{cases} 0 \\ 1 \\ 2 \\ 3 \end{cases}$$

(a) Since X has a hypergeometric distribution.

$$P\{X = k\} = \frac{\binom{M}{k}\binom{N-M}{n-k}}{\binom{N}{n}}$$

So P{A lottery ticket has 4 numbers that are the same as 4 of the 6 winning numbers}

$$= P\{X = 4\} = \frac{\binom{6}{4}\binom{42}{2}}{\binom{48}{6}} \approx 0.00105$$

(b) Since X and Y are not independent and obey different hypergeometric distributions, so

P{A lottery ticket has 5 numbers that are the same as 5 of the 6 winning numbers and one number that is the same as one of the additional 3 winning numbers}

$$= P\{\{X = 5\} \cap \{Y = 1\}\} = P\{X = 5\} \cdot P[\{Y = 1\} \mid \{X = 5\}]$$

$$= \frac{\binom{6}{5}\binom{42}{1}}{\binom{48}{6}} \cdot \frac{\binom{1}{1}\binom{41}{2}}{\binom{42}{3}} \approx 1.47 \times 10^{-6}$$

Example 2. A buyer wants to purchase a package of N = 12 electronic components. He randomly selects $n = 4$ components from the package. If these 4 components are all good, he will buy this package. If the number of packages containing defective components $m_1 = 3$ accounts for 15% and the number of packages containing defective components $m_2 = 1$ accounts for 85%, what proportion of the packages is rejected by this buyer?

Solution. $P\{m_1 = 3\} = 0.15$, $P\{m_2 = 1\} = 0.85$. $\{m_1 = 3\}$, $\{m_2 = 1\}$ is a partition of the sample space.

Let A = {4 components are all good} = {buy this package}. Then, the conditional probability of A (which has a hypergeometric distribution) is

$$P\{A|m_1=3\}=\frac{\binom{9}{4}\binom{3}{0}}{\binom{12}{4}} \qquad\qquad P\{A|m_2=1\}=\frac{\binom{11}{4}\binom{1}{0}}{\binom{12}{4}}$$

So, according to the law of total probability, the probability of A is

$$P(A)=P\{\text{buy this package}\}=P\{A|m_1=3\}P\{m_1=3\}+P\{A|m_2=1\}P\{m_2=1\}$$

$$=\frac{\binom{9}{4}\binom{3}{0}}{\binom{12}{4}}\cdot 0.15+\frac{\binom{11}{4}\binom{1}{0}}{\binom{12}{4}}\cdot 0.85=\frac{499}{825}$$

Therefore, the proportion of packages rejected by buyers is

$$P(A')=1-\frac{499}{825}=\frac{326}{825}\approx 39.5\%$$

Exercise 3.2

1. In a big city, 55% of voters support one environmental proposal. Take a random sample of 15 voters from the city. Find the probability in the sample that

- **a.** exactly 10 voters support the proposal.
- **b.** at least 12 voters support the proposal.
- **c.** at most 3 voters support the proposal.

2. Sam has 4 free-throw attempts in basketball. All attempts are independent. If the probability that he succeeds at least one time is $\frac{80}{81}$, what is the success probability for each free throw?

3. The accuracy of weather forecasts in a particular place is 0.8.

a. Find the probability that exactly 4 out of 5 forecasts are accurate.

b. Find the probability that at least 4 out of 5 forecasts are accurate.

4. Six employees in an organisation work with the help of the Internet. The probability of each employee using the Internet is 0.5, and whether each employee is using the Internet is independent of each other.

a. What is the probability that at least 3 employees are using the Internet at the same time?

b. What is the number of employees required so that the probability that they all use the Internet simultaneously is greater than 0.3?

5. The probability that someone shoots an arrow and hits the bull's eye is p. He shoots m ($m > 20$) arrows at the bull's eye in succession. Find the probability that

a. he hits the bull's eye.

b. he hits the bull's eye exactly 10 times.

c. he hits the bull's eye for at least 3 times.

6. A family has 7 children. If each child has a probability of 51% of being a girl, find the probability that the family has

a. 4 boys and 3 girls.

b. more girls than boys.

7. Anne and Cassie play a game. Anne is more skilful and can win, on average, 3 games out of every 5 games played. If Anne and Cassie play 10 games, find the probability that

a. Anne wins exactly 5 of these games.

b. Anne will win more games than Cassie.

8. How many throws of a fair die are required such that the probability of obtaining one 5 is at least 80%?

9. For a binomial variable X with $n = 10$, what is the smallest p-value such that $P\{X \geq 4\} \geq 0.95$?

10. The company manager will submit a proposal to the directors' representatives for approval. The proposal will be approved if most directors' representatives vote in favour. The manager estimates that the probability that each representative will vote in favour of the proposal is 0.6, and each representative will vote independently of the other. To have a higher probability of passing the proposal, is it better for the manager to invite 3 director representatives or 5 for the proposal meeting?

11. Having observed that 5% of the persons making reservations on a flight do not show up for the flight, an airline company sells 52 seats on a plane with 50 seats. What is the probability that there will not be a seat for every person who shows up for the flight?

12. A discrete random variable X follows a Poisson distribution with parameter λ. Given that $P\{X = 1\} = P\{X = 2\}$, find the probability that $X = 3$.

13. A book of 585 pages contains 43 typographical errors, randomly distributed throughout the book. Suppose that X, the number of errors on a randomly selected page, has a Poisson distribution. What is the probability that 10 pages chosen at random will be free of errors?

14. A system consists of 1000 parts. The failure of each part is independent of each other, and the probability of failure within a month is 0.001. Use the Poisson distribution to find the probability that the system operates normally (that is, no parts fail) within a month.

15. According to records, the number X of televisions sold in a shop each month obeys a Poisson distribution with parameter $\lambda = 7$. How many televisions should be stocked at the beginning of the month to satisfy customer demand for televisions with a probability of 0.999?

16. Assume that the number of customers entering a store per hour, X, follows a Poisson distribution with parameter $\lambda = 200$. The probability that each customer buys some products is 0.05, and whether each customer buys some products is independent of other customers. Find the probability that at least 6 customers buy some product at the store within an hour.

17. A batch of 20 items contains five defectives. Find the probability that a sample of four will contain more than one defective.

18. A school's Student Council has 18 members. Ten are in favour of Proposal A, and eight are against it. A vote is taken to determine the majority's will. If five random members of the 18 do not vote, find the probability that those opposing the proposal will win.

3.3. Cumulative distribution function of random variables, Probability density of continuous random variables

Below, the concept of the cumulative distribution function of random variables is introduced.

Definition

If X is a random variable and the endpoint x is any real number, then

$$F(x) = P\{X \le x\} \qquad \text{endpoint} \qquad x \in R$$

is the cumulative distribution function of X.

The inequality events $\{X \le x\}, \{x_1 < X \le x_2\}$ etc. are all sum events. F(x) is the probability of the sum event $\{X \le x\}$. It is an ordinary function. We can use it to study random variables using mathematical analysis methods.

If X is treated as a random point on the real number line, as shown in the following,

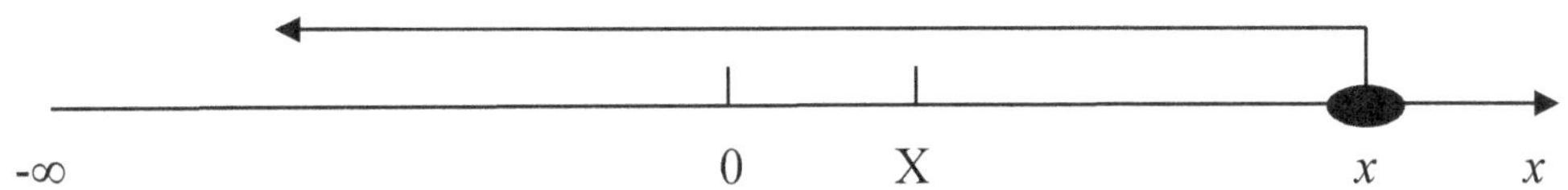

then the function value of the cumulative distribution function F(x) at the endpoint x represents the probability that X falls on the interval]-∞, x], that is, $P\{X \le x\}$.

Cumulative distribution functions have the following two basic properties:

1. For any real numbers x_1, x_2 $(x_1 < x_2)$, as shown in Graph 3.1,

$$\{X \le x_1\} \cup \{x_1 < X \le x_2\} = \{X \le x_2\}$$

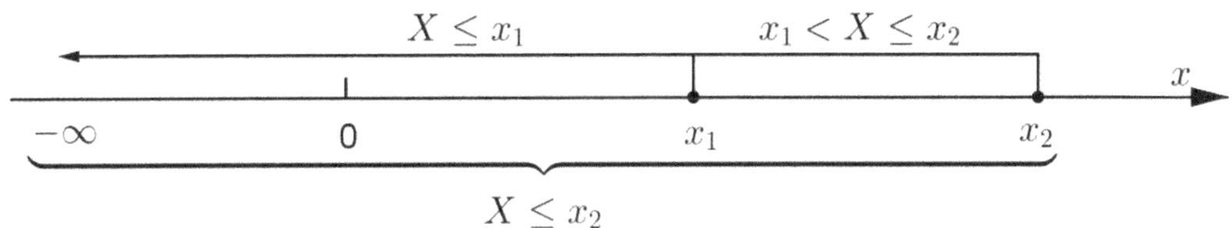

Graph 3.1

So $$P\{X \le x_1\} + P\{x_1 < X \le x_2\} = P\{X \le x_2\}$$

$$P\{x_1 < X \le x_2\} = P\{X \le x_2\} - P\{X \le x_1\} = F(x_2) - F(x_1)$$

That is, $$F(x_2) - F(x_1) = P\{x_1 < X \le x_2\} \ge 0 \qquad (x_1 < x_2)$$

Therefore, F(x) is an increasing function.

2. $0 \le F(x) \le 1$

$$F(-\infty) = \lim_{x \to -\infty} F(x) = P\{X \le -\infty\} = 0$$

$$F(\infty) = \lim_{x \to \infty} F(x) = P\{X \le \infty\} = 1$$

Proof. Because $X \in R$, then $\{X \le -\infty\}$ is an impossible event.

Hence $$F(-\infty) = \lim_{x \to -\infty} F(x) = P\{X \le -\infty\} = 0$$

Similarly, because $X \in R$, then $\{-\infty < X < \infty\}$ is a definite event.

So $$P\{X \le \infty\} = P\{X < \infty\} + P\{X = \infty\} = 1 + 0 = 1$$

Therefore, we have

$$F(\infty) = \lim_{x \to \infty} F(x) = P\{X \le \infty\} = 1$$

where $\{X = \infty\}$ is an impossible event.

3.3.1. Cumulative distribution function of discrete random variables

Let the probability distribution of the discrete random variable X be

$$P\{X = x_k\} = p_k \qquad k = 1, 2, \ldots$$

We can get the cumulative distribution function of X according to the property that probabilities can be added and listed.

$$F(x) = P\{X \le x\} = \sum_{x_k \le x} P\{X = x_k\} = \sum_{x_k \le x} p_k$$

Here, add up all p_k that satisfy $x_k \le x$. The cumulative distribution function $F(x)$ jumps at the point $x = x_k$ $(k = 1, 2, \ldots)$ with the value $p_k = P\{X = x_k\}$.

Example. The distribution of a discrete random variable X is

X	1	2	3
p_k	0.25	0.5	0.25

Find the cumulative distribution function $F(x)$ of X.

Solution. Because the cumulative distribution function of discrete X is

$$F(x) = P\{X \le x\} = \sum_{x_k \le x} P\{X = x_k\} = \sum_{x_k \le x} p_k$$

so the required cumulative distribution function is

$$F(x) = \begin{cases} 0 & x < 1 \\ 0.25 & 1 \le x < 2 \\ 0.25 + 0.5 & 2 \le x < 3 \\ 0.25 + 0.5 + 0.25 & x \ge 3 \end{cases}$$

That is,

$$F(x) = \begin{cases} 0 & x < 1 \\ 0.25 & 1 \le x < 2 \\ 0.75 & 2 \le x < 3 \\ 1 & x \ge 3 \end{cases}$$

The graph of $F(x)$ is shown in Graph 3.2, and it is a stepped curve. The jump points are at $x = 1, 2, 3$, and the jump values are 0.25, 0.5 and 0.25, respectively.

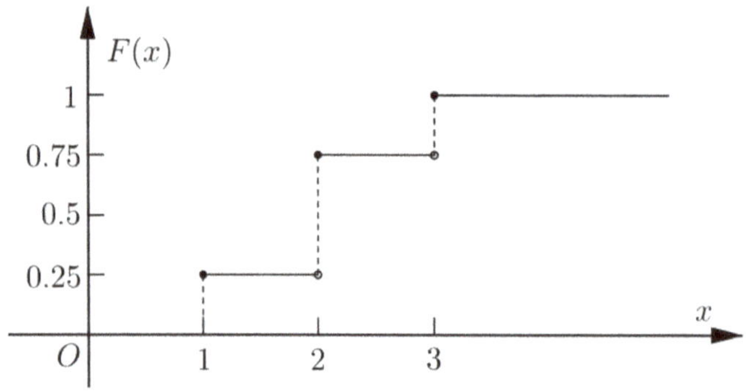

Graph 3.2

3.3.2. Cumulative distribution function of continuous random variables

Let us first read an example. From this, we can derive the formula of the cumulative distribution function F(x) of continuous random variables X.

Example. The radius of a circular target is 2 m. Assume that the probability of hitting a point on any concentric circle of the target is proportional to the area of the concentric circle. Meanwhile, all shots hit the target. Let X represent the distance between the hit point and the centre of the target. Find the cumulative distribution function of X.

Solution. Draw a picture based on the given information. From Graph 3.3, we can see that if $x < 0$, then $\{X \leq x\}$ is an impossible event.

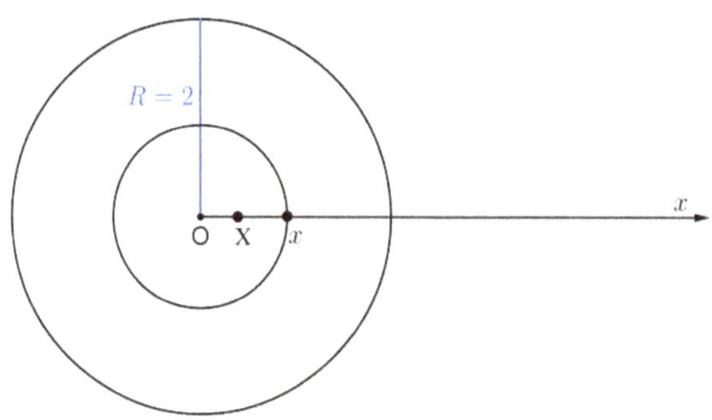

Graph 3.3

So $\quad F(x) = P\{X \le x\} = 0 \quad (x < 0)$

If $0 \le x \le 2$, according to the given information

$$P\{0 \le X \le x\} = m\pi x^2 = kx^2$$

where m is the proportional coefficient, $k = m\pi$ = constant.

We want to find the value of k. Choose $x = 2$.

Because $\quad 1 = P\{0 \le X \le 2\} = k\cdot 2^2 \quad$ so $\quad k = \dfrac{1}{4}$

Hence $\quad P\{0 \le X \le x\} = kx^2 = \dfrac{1}{4}x^2$

So $\quad F(x) = P\{X \le x\} = P\{X < 0\} + P\{0 \le X \le x\} = \dfrac{1}{4}x^2 \quad (0 \le x < 2)$

If $x \ge 2$, then according to the given information, $\{X \le x\}$ is a definite event. So

$$F(x) = P\{X \le x\} = 1 \qquad (x \ge 2)$$

Hence, we get the cumulative distribution function $F(x)$ of X.

$$F(x) = P\{X \le x\} = \begin{cases} 0 & x < 0 \\ \dfrac{x^2}{4} & 0 \le x < 2 \\ 1 & x \ge 2 \end{cases}$$

Graph of $F(x)$

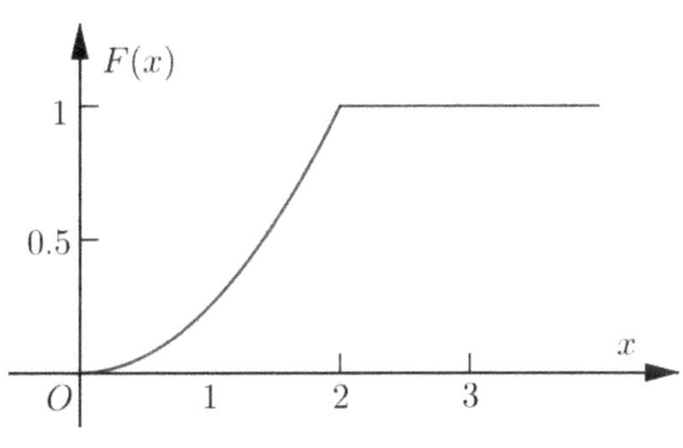

Graph 3.4

Graph 3.4 shows that the graph of F(x) is a continuous curve, so the cumulative distribution function F(x) is also a continuous function.

It is easy to observe that the cumulative distribution function F(x) can be written as an integral equation.

$$F(x) - F(-\infty) = \int_{-\infty}^{x} dF(x) = \int_{-\infty}^{x} f(x)dx$$

That is, $$F(x) = \int_{-\infty}^{x} f(x)dx \quad \text{where} \quad f(x) = \frac{dF(x)}{dx} = \begin{cases} \frac{x}{2} & 0 < x < 2 \\ 0 & \textit{elsewhere} \end{cases}$$

In other words, F(x) is the integral of function $f(x)$ on the interval]-∞, x]. At this time, X is called a continuous random variable, and its cumulative distribution function F(x) is

$$F(x) = \int_{-\infty}^{x} f(x)dx \qquad f(x) \geq 0$$

The cumulative distribution function F(x) of a continuous random variable is a continuous and increasing function.

3.3.3. Probability density of continuous random variables

Definition

If, for the cumulative distribution function F(x) of a random variable X, there exists a function $f(x)$ ($f(x) \geq 0$) such that to any real number x, we have

$$F(x) = \int_{-\infty}^{x} f(x)dx$$

then, we call X a continuous random variable, and the function $f(x)$ is called the probability density function of X, or simply the probability density.

From the definition, we know that the probability density $f(x)$ has the following properties:

1. $f(x) \geq 0 \qquad x \in \mathrm{R}$;
2. $\int_{-\infty}^{\infty} f(x)dx = \mathrm{F}(\infty) - \mathrm{F}(-\infty) = 1$;
3. $\mathrm{P}\{x_1 < \mathrm{X} \leq x_2\} = \mathrm{F}(x_2) - \mathrm{F}(x_1) = \int_{x_1}^{x_2} f(x)dx \qquad (x_1 \leq x_2)$;
4. If $f(x)$ is continuous at point x, then we have $\mathrm{F}'(x) = \dfrac{d\mathrm{F}}{dx} = f(x)$.

From Property 2, we know that the area between the curve $y = f(x)$ and the x-axis is 1, as shown in Graph 3.5.

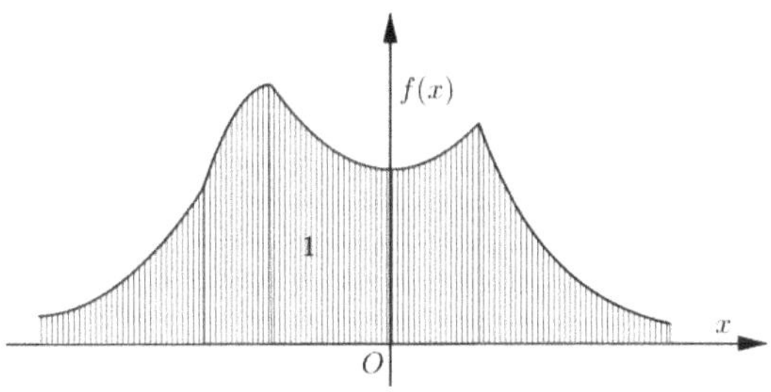

Graph 3.5

From property 3, we know that the probability that X falls on the interval $]x_1, x_2]$, that is, $\mathrm{P}\{x_1 < \mathrm{X} \leq x_2\}$, is equal to the area bounded by the curve $y = f(x)$, the x-axis, $x = x_1$ and $x = x_2$, as shown in Graph 3.6.

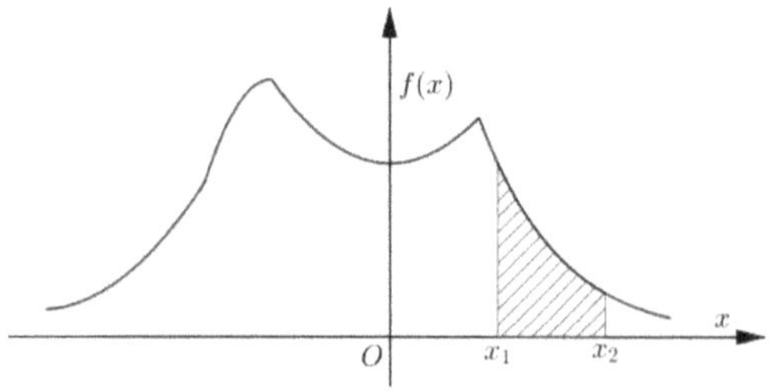

Graph 3.6

$$P\{x_1 < X \le x_2\} = \int_{x_1}^{x_2} f(x)dx = F(x_2) - F(x_1)$$

According to property 4, if $f(x)$ is continuous at point x, then we have

$$f(x) = F'(x) = \frac{dF}{dx}$$

That is, $$f(x) = \frac{dF}{dx}$$

Here, we can observe that the definition of probability density $f(x)$ is similar to the definition of mass linear density $p = \dfrac{dm}{dx}$ in physics. This is why $f(x)$ is called a probability density.

Moreover, since

$$f(x) = \frac{dF}{dx} = \lim_{\Delta x \to 0^+} \frac{\Delta F}{\Delta x} = \lim_{\Delta x \to 0^+} \frac{F(x+\Delta x) - F(x)}{\Delta x} = \lim_{\Delta x \to 0^+} \frac{P\{x \le X \le x + \Delta x\}}{\Delta x}$$

$$= \frac{\lim_{\Delta x \to 0^+} P\{x \le X \le x + \Delta x\}}{\lim_{\Delta x \to 0^+} \Delta x} = \frac{P\{x \le X \le x\}}{\lim_{\Delta x \to 0^+} \Delta x} = \frac{P\{X = x\}}{\lim_{\Delta x \to 0^+} \Delta x}$$

hence $$P\{X = x\} = f(x) \cdot \lim_{\Delta x \to 0^+} \Delta x$$

From $\lim_{\Delta x \to 0^+} \Delta x = 0$, we get

$$P\{X = x\} = 0$$

We can get the same result using a calculation formula or analytical method.

$$P\{X = x\} = P\{x \le X \le x\} = \int_x^x f(x)dx = F(x) - F(x) = 0$$

Or $$P\{X = x\} = \frac{\binom{f}{1}}{\binom{\infty}{1}} = 0$$

It can be seen from the above that the comprehensive and intuitive expression for the probability of the sum event is

$$P\{a \le X \le b\} = \sum_{i=1}^{\infty} P\{X = x_i\} = \sum_{i=1}^{\infty} f(x_i) \lim_{\Delta x_i \to 0} \Delta x_i = \int_a^b f(x)dx = F(b) - F(a)$$

Let $a = x_1$, $b = x_2$. Then, the calculation formula for probability is

$$P\{x_1 \le X \le x_2\} = \int_{x_1}^{x_2} f(x)dx = F(x_2) - F(x_1)$$

It can also be seen that for a continuous random variable X, the probability of taking any specified single real number x is 0, that is,

$$P\{X = x\} = 0$$

According to this, when we calculate the probability that a continuous random variable X falls into a certain interval, there is no need to distinguish whether the interval is an open interval, a closed interval, or a semi-open and semi-closed interval, that is,

$$P\{x_1 \le X \le x_2\} = P\{x_1 < X \le x_2\} = P\{x_1 \le X < x_2\} = P\{x_1 < X < x_2\}$$

Proof. Since $P\{x_1 \le X \le x_2\} = P\{x_1 < X \le x_2\} + P\{X = x_1\} = P\{x_1 < X \le x_2\} + 0$

Likewise, we can have

$$P\{x_1 \le X \le x_2\} = P\{x_1 < X < x_2\} - P\{x_1 < X < x_2\}$$

Therefore $\qquad P\{x_1 \le X \le x_2\} = P\{x_1 < X \le x_2\} = P\{x_1 \le X < x_2\} = P\{x_1 < X < x_2\}$ □

Note:

Here, the event $\{X = x\} \neq \varnothing$ and it is not an impossible event. $P\{X = x\} = 0$ simply means that the probability of event $\{X = x\}$ occurring is 0. That is, if $A = \varnothing$, then we have $\qquad P(A) = P(\varnothing) = 0$

But, if P(A) = 0, then this does not mean that A is an impossible event.

Exercise 3.3

1. The probability distribution of a discrete random variable is shown in the following table.

X	-2	1	3
p	$\frac{1}{5}$	$\frac{1}{2}$	$\frac{3}{10}$

a. Find the cumulative distribution function $F(x)$.

b. Find the following probabilities.

i. $P\{X \le 0\}$ **ii.** $P\{\frac{1}{4} < X \le \frac{5}{4}\}$ **iii.** $P\{1 \le X \le 3\}$

2. The cumulative distribution function $F(x)$ of a discrete random variable X is

$$F(x)=\begin{cases}0 & x<-3\\ 0.3 & -3\le x<-2\\ 0.5 & -2\le x<3\\ 1 & x\ge 3\end{cases}$$

Find the probability distribution of X.

3. Which of the following functions are probability density functions?

a. $f(x)=\begin{cases}x(2-x) & 0<x<2\\ 0 & elsewhere\end{cases}$

b. $f(x)=\begin{cases}\dfrac{1}{\lambda}e^{-\frac{x-\theta}{\lambda}} & x>\theta\\ 0 & elsewhere\end{cases}\qquad \lambda>0$

c. $f(x)=\begin{cases}\sin x & 0<x<\dfrac{\pi}{2}\\ 0 & elsewhere\end{cases}$

d. $f(x)=\dfrac{1}{2}e^{-|x|}\qquad -\infty<x<\infty$

e. $f(x)=\begin{cases}0 & x<0\\ \dfrac{x+1}{9} & 0\le x<1\\ \dfrac{2(2x-1)}{9} & 1\le x<\dfrac{3}{2}\\ \dfrac{2(5-2x)}{9} & \dfrac{3}{2}\le x<2\\ \dfrac{4}{27} & 2\le x<5\\ 0 & elsewhere\end{cases}$

4. Let $k>0$ be a constant and

$$f(x)=\begin{cases}kx(1-x) & 0<x<1\\ 0 & elsewhere\end{cases}$$

a. Find the value of k if $f(x)$ is a probability density function.

b. Find $P\{X>0.2\}$.

5. The amount of bread (in hundreds of kilos) that a bakery sells in a day is a random variable with the probability density

$$f(x)=\begin{cases} cx & 0\le x<3 \\ c(6-x) & 3\le x<6 \\ 0 & elsewhere \end{cases}$$

a. Find the value of c.

b. What is the probability that the number of kilos of bread sold in a day is

i. more than 300 kilos?

ii. between 150 and 450 kilos?

6. The diameter X of an electric cable is assumed to be a continuous random variable with a probability density function

$$f(x)=\begin{cases} 6x(1-x) & 0\le x\le 1 \\ 0 & elsewhere \end{cases}$$

Find a number k such that $P\{X<k\}=2P\{X>k\}$.

7. The probability density function of X is

$$f(x)=\begin{cases} ax & 1<x<2 \\ b & 2\le x<3 \\ 0 & elsewhere \end{cases}$$

If $P\{1<X<2\}=P\{2<X<3\}$, find the values of a and b.

8. The probability density function of X is

$$f(x)=\frac{a}{1+x^2} \qquad -\infty<x<\infty$$

Find

a. the value of constant a.

b. the probability of $P\{|X|<1\}$.

9. The probability density function of X is

$$f(x)=\begin{cases} a+x & -1\le x\le 0 \\ a-x & 0<x<1 \\ 0 & elsewhere \end{cases}$$

Find

a. the value of constant a.

b. the probability of $P\{|X| < \frac{1}{2}\}$.

3.4. Three important continuous random variables

3.4.1. Exponential distribution

If the probability density of a continuous random variable X is

$$f(x)=\begin{cases} \lambda e^{-\lambda x} & x>0 \\ 0 & elsewhere \end{cases}$$ where $\lambda > 0$ and λ is a constant.

then we say that X has an exponential distribution with parameter λ. The graph of $f(x)$ is shown in Graph 3.7.

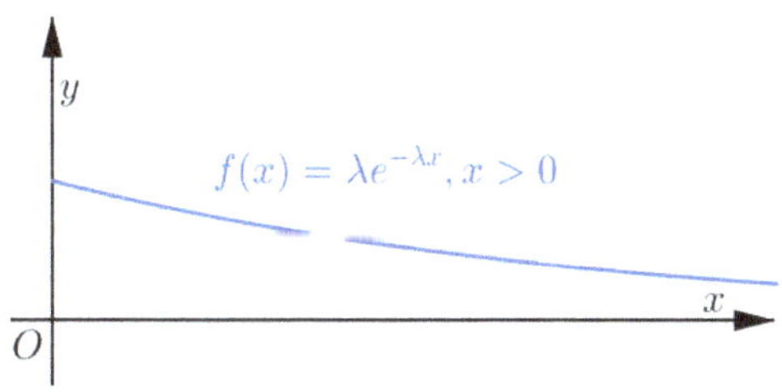

Graph 3.7

Example. The probability density $f(x)$ of a random variable X is

$$f(x)=\begin{cases} ke^{-3x} & x>0 \\ 0 & x\le 0 \end{cases}$$

Find the constant k and $P\{X > 0.1\}$.

Solution. Because $\int_{-\infty}^{\infty} f(x)dx = 1$

$$\int_{-\infty}^{\infty} f(x)dx = \int_{-\infty}^{0} 0dx + \int_{0}^{\infty} ke^{-3x}dx = \int_{0}^{\infty} ke^{-3x}dx = \left[k(-\frac{1}{3}e^{-3x}) \right]_0^{\infty} = \frac{k}{3}$$

That is, $\frac{k}{3} = 1$ so $k = 3$

So, the probability density $f(x)$ of X is

$$f(x) = \begin{cases} 3e^{-3x} & x > 0 \\ 0 & x \leq 0 \end{cases}$$

$$P\{X > 0.1\} = \int_{0.1}^{\infty} f(x)dx = \int_{0.1}^{\infty} 3e^{-3x}dx = \left[-e^{-3x} \right]_{0.1}^{\infty} \approx 0.741$$

3.4.2. Uniform distribution

If the probability density of a continuous random variable X is

$$f(x) = \begin{cases} \dfrac{1}{b-a} & a \leq x \leq b \\ 0 & elsewhere \end{cases}$$

then we say that on the interval $[a, b]$, X has a uniform distribution.

Uniform distribution has the significance that probability distributes equally because

$$P\{x \leq X \leq x+L\} = \int_{x}^{x+L} f(x)dx = \int_{x}^{x+L} \frac{1}{b-a}dx = \left[\frac{1}{b-a}x \right]_x^{x+L} = \frac{L}{b-a}$$

It can be seen from the probability result $\frac{L}{b-a}$ that the probability that X falls on any subinterval $[x, x+L]$ with the same length L on the interval $[a, b]$ is equal. In other words, the probability only depends on the length of the subinterval, not its position.

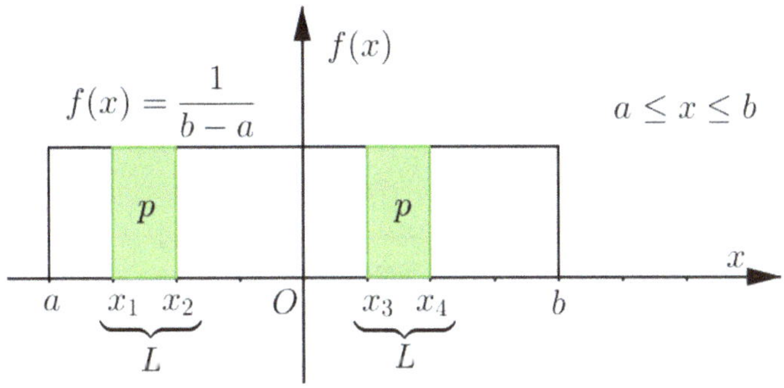

Graph 3.8

From Graph 3.8, we can see that although the positions of the subintervals $[x_1, x_2]$ and $[x_3, x_4]$ are different, their lengths are equal. That is,

$$P\{ x_1 \le X \le x_1 + L\} = P\{ x_3 \le X \le x_3 + L\} = \frac{L}{b-a}$$

$$(x_2 = x_1 + L, \quad x_4 = x_3 + L)$$

So uniform distribution is a distribution where the probability distributes uniformly.

Example 1. Random variable X has a uniform distribution on the interval $]a, b[$ and its probability density is

$$f(x) = \begin{cases} \dfrac{1}{b-a} & a < x < b \\ 0 & elsewhere \end{cases}$$

Construct the cumulative distribution function F(x) of X.

Solution. Because the cumulative distribution function of X is

$$F(x) = P\{X \le x\} = \int_{-\infty}^{x} f(x)dx$$

so the required cumulative distribution function is

$$F(x) = \begin{cases} \int_{-\infty}^{x} f(x)dx = \int_{-\infty}^{x} 0dx = 0 & x < a \\ \int_{-\infty}^{x} f(x)dx = \int_{-\infty}^{a} 0dx + \int_{a}^{x} \frac{1}{b-a}dx = \frac{x-a}{b-a} & a \le x < b \\ \int_{-\infty}^{x} f(x)dx = \int_{-\infty}^{a} 0dx + \int_{a}^{b} \frac{1}{b-a}dx + \int_{b}^{x} 0dx = 1 & x \ge b \end{cases}$$

That is,

$$F(x) = \begin{cases} 0 & x < a \\ \frac{x-a}{b-a} & a \le x < b \\ 1 & x \ge b \end{cases}$$

The graph of F(x) is shown in Graph 3.9.

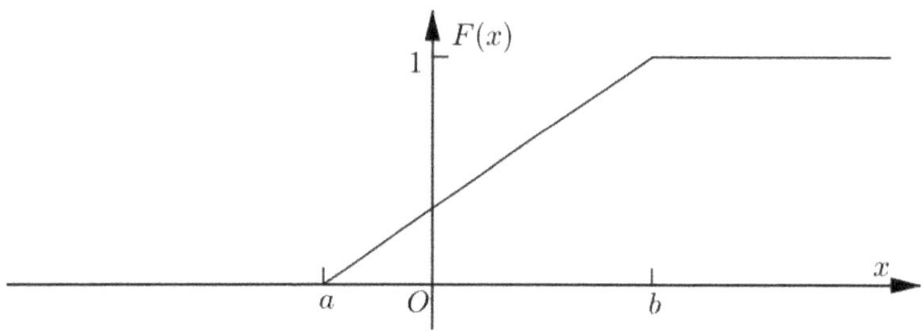

Graph 3.9

The graph of F(x) is a continuous curve.

Example 2. Let the value of resistance R be a random variable that distributes uniformly on the interval 900 Ω ~1100 Ω. Calculate the probability density of R and the probability that R takes a value from the interval 950 Ω ~1050 Ω.

Solution. According to the given information, the probability density of R is

$$f(r) = \begin{cases} \frac{1}{1100-900} & 900 < r < 1100 \\ 0 & elsewhere \end{cases}$$

So $$P\{950 \le R \le 1050\} = \int_{950}^{1050} \frac{1}{200} dr = 0.5$$

3.4.3. Normal distribution

If the probability density of a continuous random variable X is

$$f(x) = \frac{1}{\sigma\sqrt{2\pi}} e^{-\frac{1}{2}\left(\frac{x-\mu}{\sigma}\right)^2} \qquad -\infty < x < \infty$$

where μ, σ ($\sigma > 0$) are constants, then we say that X is normally distributed with parameters μ and σ, or X has the Gauss distribution, denoted as $X \sim N(\mu, \sigma^2)$. The parameter μ is the expectation and σ is the standard deviation. That is, $\mu = E(X)$, $\sigma^2 = \text{Var}(X)$. We will learn their meaning later.

The graph of $f(x)$ is shown in Graph 3.10.

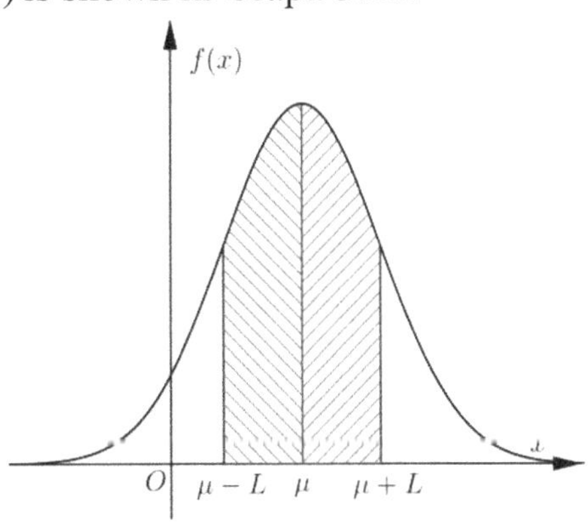

Graph 3.10

The probability density $f(x)$ of the normal distribution has the following properties:

(1) The graph of $f(x)$ is symmetric about the line $x = \mu$. This means that for any $L > 0$, we have

$$P\{\mu - L \le X \le \mu\} = P\{\mu \le X \le \mu + L\}$$

(2) When $x = \mu$, $f(x)$ takes the maximum value.

$$f(x)_{\max} = f(\mu) = \frac{1}{\sigma\sqrt{2\pi}}$$

The farther x is from μ, the smaller the value of $f(x)$ is. The x-axis is the horizontal asymptote of the curve of $f(x)$.

(3) If σ is kept constant and the value of μ is changed, the graph moves parallel to the x-axis, but the shape does not change, as shown in Graph 3.11. Therefore, the position of the probability density $y = f(x)$ of the normal distribution is completely determined by the parameter μ. μ is called a positional parameter.

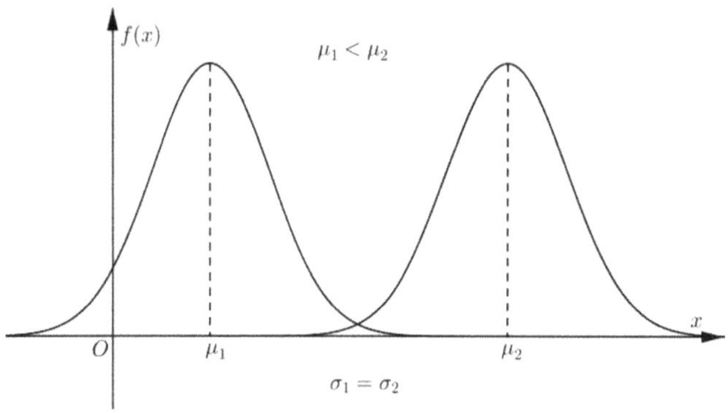

Graph 3.11

(4) If μ is kept constant and the value of σ is changed, since the maximum value of $f(x)$ is $f(\mu) = \dfrac{1}{\sigma\sqrt{2\pi}}$, then the smaller the value of σ, the sharper of the graph of $f(x)$ and the higher probable that X takes values around μ, as shown in Graph 3.12.

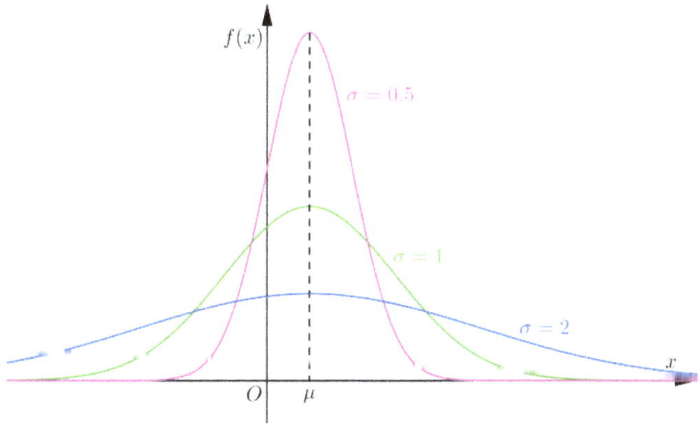

Graph 3.12

X ~ N (μ, σ^2), and the cumulative distribution function of X is

$$F(x) = P\{X \le x\} = \int_{-\infty}^{x} f(x)dx = \frac{1}{\sigma\sqrt{2\pi}} \int_{-\infty}^{x} e^{-\frac{1}{2}\left(\frac{x-\mu}{\sigma}\right)^2} dx$$

The graph of F(x) is shown in Graph 3.13.

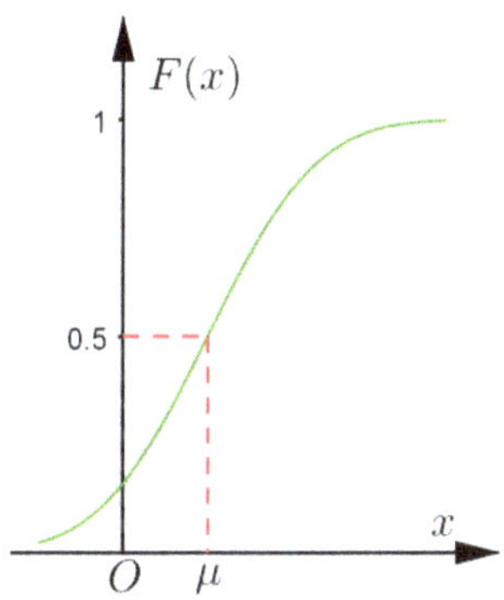

Graph 3.13

In particular, when $\mu = 0$, $\sigma = 1$, we say that X has the standard normal distribution, that is, X ~ N (0, 1). Its probability density is denoted by the letter φ (φ is the lower case Greek letter f and is pronounced phi), and its cumulative distribution function is denoted by the letter Φ (upper case Greek letter phi).

So we have

$$\varphi(x) = \frac{1}{\sqrt{2\pi}} e^{-\frac{x^2}{2}} \qquad \Phi(x) = \frac{1}{\sqrt{2\pi}} \int_{-\infty}^{x} e^{-\frac{x^2}{2}} dx$$

It is easy to know that

$$\Phi(-x) = 1 - \Phi(x)$$

The values of $\Phi(x)$ can be obtained using a mathematical table or a calculator.

The graph of $\varphi(x)$ of X ~ N (0, 1) is shown in Graph 3.14.

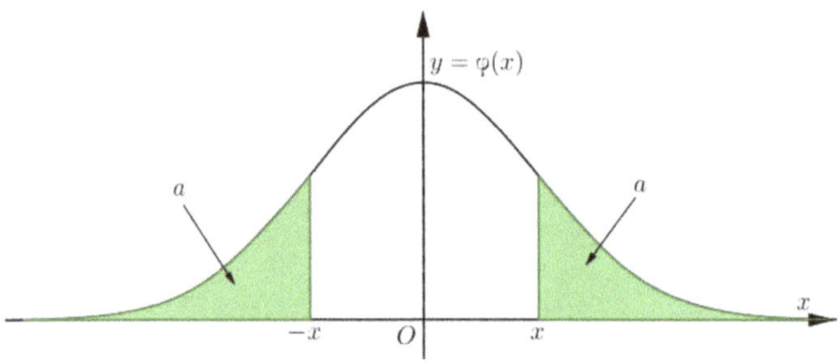

Graph 3.14

In the following, we use the graph of $\varphi(x)$ to prove

$$\Phi(-x) = 1 - \Phi(x).$$

Proof. Since $\Phi(-x) = P\{X \le -x\} = \int_{-\infty}^{-x} f(x)dx = a$

$$\Phi(x) = P\{X \le x\} = \int_{-\infty}^{x} f(x)dx = 1 - a = 1 - \Phi(-x)$$

so $\Phi(-x) = 1 - \Phi(x)$ □

Theorem: If X ~ N (μ, σ^2), we can use linear transformation to change it to the standard normal distribution N (0, 1).

That is,

$$X \sim N(\mu, \sigma^2) \xLeftrightarrow{\text{linear transformation}} \frac{X-\mu}{\sigma} \sim N(0,1)$$

Let $\mathbf{Z} = \dfrac{X-\mu}{\sigma}$ **Z is called a standardised random variable.**

Proof. Because X ~ N (μ, σ^2) for any interval $[x_1, x_2]$ we have

$$P\{x_1 \le X \le x_2\} = \frac{1}{\sigma\sqrt{2\pi}} \int_{x_1}^{x_2} e^{-\frac{1}{2}\left(\frac{x-\mu}{\sigma}\right)^2} dx$$

Let $z = \dfrac{x-\mu}{\sigma}$ so $dz = \dfrac{dx}{\sigma}$ $dx = \sigma dz$

When $x = x_1$, $z_1 = \dfrac{x_1-\mu}{\sigma}$; When $x = x_2$, $z_2 = \dfrac{x_2-\mu}{\sigma}$.

So $$\frac{1}{\sigma\sqrt{2\pi}} \int_{x_1}^{x_2} e^{-\frac{1}{2}\left(\frac{x-\mu}{\sigma}\right)^2} dx = \frac{1}{\sigma\sqrt{2\pi}} \int_{\frac{x_1-\mu}{\sigma}}^{\frac{x_2-\mu}{\sigma}} e^{-\frac{1}{2}z^2} \sigma dz = \int_{\frac{x_1-\mu}{\sigma}}^{\frac{x_2-\mu}{\sigma}} [\frac{1}{\sqrt{2\pi}} e^{-\frac{1}{2}z^2}] dz$$

(the bracketed integrand → N (0, 1))

Therefore $$P\{x_1 \le X \le x_2\} = \frac{1}{\sqrt{2\pi}} \int_{\frac{x_1-\mu}{\sigma}}^{\frac{x_2-\mu}{\sigma}} e^{-\frac{1}{2}z^2} dz = \Phi(\frac{x_2-\mu}{\sigma}) - \Phi(\frac{x_1-\mu}{\sigma})$$

$$= P\{\frac{x_1-\mu}{\sigma} \le Z \le \frac{x_2-\mu}{\sigma}\}$$

(Z → N (0, 1))

By comparison, we can get

$$Z = \frac{X-\mu}{\sigma} \sim N(0, 1)$$

Hence

$$X \sim N(\mu, \sigma^2) \xLeftrightarrow{\text{linear transformation}} \frac{X-\mu}{\sigma} \sim N(0,1)$$ □

The above theorem can be used to calculate the probability of a normally distributed random variable X.

$$P\{x_1 \le X \le x_2\} = P\{\frac{x_1-\mu}{\sigma} \le \frac{X-\mu}{\sigma} \le \frac{x_2-\mu}{\sigma}\} = \Phi(\frac{x_2-\mu}{\sigma}) - \Phi(\frac{x_1-\mu}{\sigma})$$

That is, $P\{x_1 \le X \le x_2\} = \Phi(\frac{x_2-\mu}{\sigma}) - \Phi(\frac{x_1-\mu}{\sigma})$

So, it is very easy to calculate using this formula. The value of $\Phi(x)$ can be found in a table or a calculator. We can calculate $\Phi(-x)$ using the relation $\Phi(-x) = 1 - \Phi(x)$.

Example 1. In a city, the heights of high school boys are roughly normally distributed, with a mean of 175 and a variance of 25. What is the probability that the height of a randomly selected boy is

(a) more than 180 cm;

(b) between 165 cm and 170 cm;

(c) There are 260 boys in a high school. How many boys' heights are in the intervals of (a) and (b)?

Solution. Let the height of a boy be X. $X \sim N(175, 25)$ $\mu=175$, $\sigma = \sqrt{25} = 5$

(a) Since $P\{x_1 \le X \le x_2\} = \Phi(\frac{x_2-\mu}{\sigma}) - \Phi(\frac{x_1-\mu}{\sigma})$

so

$$\begin{aligned} P\{X > 180\} &= P\{180 < X < \infty\} \\ &= \Phi(\frac{\infty-175}{5}) - \Phi(\frac{180-175}{5}) \\ &= \Phi(\infty) - \Phi(1) \approx 1 - 0.841 = 0.159 \end{aligned}$$

(b) $P\{165 \le X \le 170\} = \Phi(\frac{170-175}{5}) - \Phi(\frac{165-175}{5}) = \Phi(-1) - \Phi(-2)$

Since $\Phi(-x) = 1 - \Phi(x)$

so $$P\{165 \le X \le 170\} = [1 - \Phi(1)] - [1 - \Phi(2)]$$

$$= \Phi(2) - \Phi(1) \approx 0.977 - 0.841 = 0.136$$

(c) (i) There are $n = 260$ boys in the high school. Let the number of boys with a height over 180 cm be k. Then

$$k = n \cdot P\{X > 180\} = 260 \cdot 0.159 \approx 41$$

(ii) Similarly, let the number of boys whose height is between 165 cm and 170 cm be m. Then

$$m = n \cdot P\{165 \le X \le 170\} = 260 \cdot 0.136 \approx 35$$

Example 2. Let $X \sim N(\mu, 100)$. Find the value of μ such that $P\{X \le 80\} = 0.85$.

Solution. Since $P\{x_1 \le X \le x_2\} = \Phi(\frac{x_2 - \mu}{\sigma}) - \Phi(\frac{x_1 - \mu}{\sigma})$ $\quad \sigma = \sqrt{100} = 10$

so $$P\{X \le 80\} = P\{-\infty < X \le 80\}$$

$$= \Phi(\frac{80 - \mu}{10}) - \Phi(\frac{-\infty - \mu}{10})$$

$$= \Phi(\frac{80 - \mu}{10}) - \Phi(-\infty)$$

$$= \Phi(\frac{80 - \mu}{10}) - 0$$

That is, $$P\{X \le 80\} = \Phi(\frac{80 - \mu}{10}) = 0.85$$

From a table or calculator, we find

$$\Phi(1.04) = 0.85$$

So $\frac{80 - \mu}{10} = 1.04$ Therefore $\mu = 69.6$

Example 3. An exam is conducted. If the scores obtained by all candidates approximately follow a normal distribution, then the exam is generally considered reasonable and desirable. Teachers often use test scores to estimate the normal parameters μ (average score) and σ^2 (variance) and then rate those with scores exceeding $\mu + \sigma$ as A, those with scores between μ and $\mu + \sigma$ as B, those with scores between $\mu - \sigma$ to μ as C, those with scores between $\mu - 2\sigma$ to $\mu - \sigma$ as D, and those with scores below $\mu - 2\sigma$ as F. Find the probability of each score level.

Solution. Let the student's score be X. Because

$$P\{X > \mu+\sigma\} = P\{\frac{X-\mu}{\sigma} > 1\} = 1-\Phi(1) \approx 0.159$$

$$P\{\mu < X < \mu+\sigma\} = P\{0 < \frac{X-\mu}{\sigma} < 1\} = \Phi(1)-\Phi(0) \approx 0.341$$

$$P\{\mu-\sigma < X < \mu\} = P\{-1 < \frac{X-\mu}{\sigma} < 0\} = \Phi(0)-\Phi(-1) \approx 0.341$$

$$P\{\mu-2\sigma < X < \mu-\sigma\} = P\{-2 < \frac{X-\mu}{\sigma} < -1\} = \Phi(-1)-\Phi(-2) \approx 0.136$$

$$P\{X < \mu-2\sigma\} = P\{\frac{X-\mu}{\sigma} < -2\} = \Phi(-2) \approx 0.0228$$

so, roughly speaking, in this exam, 16% got an A, 34% got a B, 34% got a C, 14% got a D, and 2% got an F. □

To facilitate future applications for standard normal random variables, we introduce the definition of a quantile a. Let X ~ N (0, 1). If z_a satisfies the condition

$$P\{X > z_a\} = a \qquad 0 < a < 1$$

then point z_a is called the upper a quantile point of the standard normal distribution, as shown in Graph 3.15.

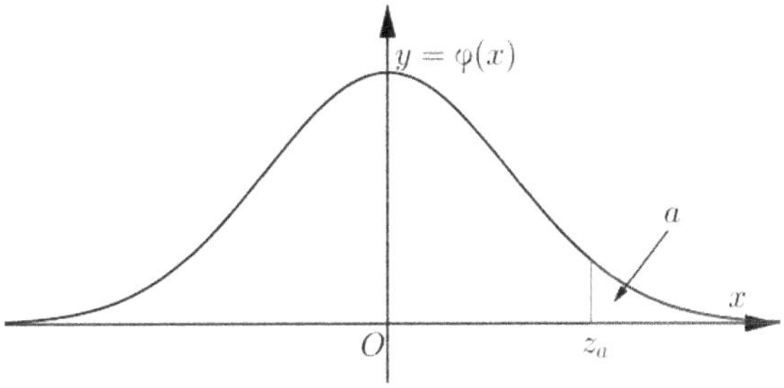

Graph 3.15

For example, from a mathematical table or calculator, we can get

$$z_{0.05} = 1.645, \qquad z_{0.005} = 2.57, \qquad z_{0.001} = 3.10$$

In natural and social phenomena, a large number of random variables obey or approximately obey normal distribution. Normal random variables are particularly important in theoretical research and practical applications of probability theory and mathematical statistics.

Note: When finding the probability of a normal distribution, if you look up a mathematical table, you need to transform it into a standard normal distribution. However, using a calculator does not necessarily require standardisation. You only need to enter μ, σ, the upper and lower limit values to get the probability of the normal distribution.

Exercise 3.4

1. Suppose that the duration in minutes of long-distance phone calls is an exponentially distributed random variable with parameter $\lambda = \frac{1}{6}$. It has the following probability density function

$$f(x) = \begin{cases} \frac{1}{6} e^{-\frac{x}{6}} & x > 0 \\ 0 & x \leq 0 \end{cases}$$

Find the probability that the duration of a long-distance phone call

a. will exceed 10 minutes.

b. will be between 7 and 9 minutes.

c. will be less than 4 minutes.

d. will be less than 10 minutes, given that it was longer than 4 minutes.

2. The time (in hours) required to repair a computer is an exponentially distributed random variable with parameter $\lambda = \frac{1}{2}$.

a. What is the probability that the repair time exceeds 2 hours?

b. If the repair time exceeds 2 hours, what is the probability that it will exceed 4 hours?

3. The number of years a particular laptop will run is an exponentially distributed random variable with parameter $\lambda = \frac{1}{4}$. Max buys a new laptop today. What is the probability that it will work after 4 years?

4. Let B be a random variable that is uniformly distributed on [0, 1]. What is the probability that the equation

$$x^2 + 4Bx + 1 = 0$$

has two distinct real roots?

5. If X ~ N (4, 9), find

a. $P\{X > 6\}$ **b.** $P\{3 < X < 8\}$ **c.** $P\{X < 2\}$ **d.** $P\{X > 5 | 3 < X < 8\}$

6. A machine produces bolts X, which are normally distributed, and $X \sim N(4, 0.09)$, where measurements are in mm. Bolts are measured accurately, and bolts smaller than 3.5 mm or bigger than 4.4 mm are rejected. Out of a batch of 5000 bolts, how many would be acceptable?

7. A normal distribution of a continuous random variable has a mean of 34 and a standard deviation of 2.8.

- **a.** Find the probability that a value selected at random will be between 26 and 38.
- **b.** Above what value do 75% of the values lie?
- **c.** Below what value do 85% of the values lie?

8. In a particular city, the number of kilos of newspaper recycled monthly by households follows a normal distribution with $\mu = 7.6$ kilos and $\sigma = 2.3$ kilos. What percentage of households recycle between 5 and 10 kilos of newspaper monthly?

9. Let X be a random variable that obeys a standard normal distribution.

- **a.** Find the probability of $P\{X < 2\}$.
- **b.** Find the probability of $P\{|X| < 2\}$.
- **c.** Given that for a constant a, $P\{|X| > a\} < 0.1$, find the value of a.

10. $X \sim N(1, 4)$.

- **a.** Find the probability of $P\{0 \le X \le 4\}$ and $P\{|X| > 2.5\}$.
- **b.** Find the value of a constant c such that $P\{X > c\} = 2P\{X \le c\}$.

11. Machines produce items. Only items weighing between 96g and 104g will be classified as salable. Find the probability of salable items if the weight of the items

- **a.** obey a uniform distribution between 95 and 105.

b. obey a normal distribution with the mean $\mu = 100$ and $\sigma = 2$.

12. $X \sim N(0, \sigma^2)$. Given that $P\{|X| \leq 20\} = 0.8$. Find the value of σ.

13. A manufacturer knows that the diameters of ball bearings he produces follow a normal distribution but does not know the mean and standard deviation. However, an inspection system rejects all bearings larger than 4.2 cm and those under 3.6 cm. Out of 1000 ball bearings, 8% are too small, and 5.3% are too big. What are the mean and standard deviation of the ball bearings produced?

14. Suppose that in a specific population, the distribution of an exam is normally distributed with mean $\mu = 600$ and standard deviation $\sigma = 80$.

a. What is the probability of a score less than 450 or greater than 750?

b. What score is equal to the 95th percentile?

c. 10 students are selected at random.

- **i.** What is the probability that 4 will score 700 or higher?
- **ii.** Given that at least 4 of them score 700 or higher, what is the probability that at most 6 of them score 700 or higher?

15. If $Z \sim N(0, 1)$, find the value of c if

a. $P\{|Z| > c\} = 0.05$.

b. $P\{|Z| < c\} = 0.9$.

16. For a normally distributed random variable X, 5% of the values are less than 65.4 and 10% of the values are greater than 69.8. Find the mean and standard deviation of X.

17. The life in hours of lighting bolts made by a factory is normally distributed with parameters $\mu = 4000$ hours and σ.

a. What is the maximum allowable value for σ (correct to two decimal places) if the probability that the life X of a bolt is between 3200 and 4800 hours is 0.7?

Use the value of σ found in a.

b. What is the probability that the life of a bolt is shorter than 2500 hours, correct to four significant figures?

c. In a package of 20 bolts, what is the probability that there are 2 or more bolts whose life is shorter than 2500 hours, correct to three significant figures?

18. Mia uses her mobile phone for X minutes each day. X is a random variable which may be modelled by a normal distribution with a mean of 45 minutes and a standard deviation of 6 minutes.

a. Find the probability that on a particular day, Mia uses her mobile phone for

i. 35 minutes.

ii. less than 35 minutes.

b. Calculate an interval, symmetrical about 45 minutes, within which X will lie on 80% of days, correct to minutes.

c. Find the probability that on 7 randomly selected days, there are at least 3 days that Mia spends at least 50 minutes on her phone.

19. The time spent by a company's employees commuting to work is normally distributed. Five per cent of the employees commute for more than 100 minutes, and 10 per cent commute for less than 30 minutes.

a. Find the mean and standard deviation of the normal distribution, correct to two decimal places.

b. 10 employees are chosen at random. Given that at least 2 of them spend between 25 and 50 minutes commuting, what is the probability that at most 5 of them spend between 25 and 50 minutes commuting?

20. The time, in minutes, required for a student to travel from home to a morning class is uniformly distributed between 20 and 25.

a. If the student leaves home promptly at 7:53, what is the probability that the student will not be late for the class at 8:15?

b. If in a 5-day week, the student always leaves home promptly at 7:53, what is the probability that the student will not be late for the class at 8:15 for at most 3 days?

3.5.* Distribution of functions of random variables

In practice, we are often more interested in some random variable's function g(X). This section will discuss how to find the distribution of its function Y = g(X) from the given distribution of the random variable X.

Example 1. Let the random variable X have the following distribution. Find the distribution of $Y = (X-1)^2$.

X	-1	0	1	2
p_k	0.2	0.3	0.1	0.4

Solution.
$$Y = (X-1)^2 = \begin{cases} (-1-1)^2 \\ (0-1)^2 \\ (1-1)^2 \\ (2-1)^2 \end{cases} = \begin{cases} 4 \\ 1 \\ 0 \\ 1 \end{cases} = \begin{cases} 0 \\ 1 \\ 4 \end{cases}$$

So
$$P\{Y=0\} = P\{(X-1)^2 = 0\} = P\{X=1\} = 0.1$$

$$P\{Y = 1\} = P\{(X-1)^2 = 1\} = P\{X = 0\} + P\{X = 2\} = 0.3 + 0.4 = 0.7$$

$$P\{Y = 4\} = P\{(X-1)^2 = 4\} = P\{X = -1\} = 0.2$$

That is, the distribution of Y is

Y	0	1	4
p_k	0.1	0.7	0.2

Example 2. Let the random variable X have a probability density

$$f_X(x) = \begin{cases} \dfrac{x}{2} & 0 < x < 2 \\ 0 & elsewhere \end{cases}$$

Find the probability density of the random variable Y = 2X + 5.

Solution. First, find the distribution function $F_Y(y)$ of Y = 2X + 5.

$$F_Y(y) = P\{Y \le y\} = P\{2X + 5 \le y\} = P\{X \le \frac{y-5}{2}\}$$

$$= \int_{-\infty}^{\frac{y-5}{2}} f_X(x)dx = F_X(\frac{y-5}{2}) - F_X(-\infty) = F_X(\frac{y-5}{2})$$

$$f_Y(y) = F_Y'(y) = \frac{dF_Y(y)}{dy} = \frac{dF_X(\frac{y-5}{2})}{dy} = \frac{dF_X(\frac{y-5}{2})}{d(\frac{y-5}{2})} \cdot \frac{d(\frac{y-5}{2})}{dy}$$

$$= f_X(\frac{y-5}{2}) \cdot \frac{1}{2} = \frac{1}{2}(\frac{y-5}{2}) \cdot \frac{1}{2} = \frac{y-5}{8} \qquad 0 < \frac{y-5}{2} < 2$$

So the probability density of Y = 2X + 5 is

$$f_Y(y) = \begin{cases} \dfrac{y-5}{8} & 5 < y < 9 \\ 0 & elsewhere \end{cases}$$

Example 3. Let the random variable X have a probability density $f_X(x)$, $-\infty < x < \infty$.

Find the probability density of $Y = X^2$.

Solution. First, find the distribution function $F_Y(y)$. Since $Y = X^2 \geq 0$, so when $y \leq 0,\ F_Y(y) = 0$.

When $y > 0$, we have

$$F_Y(y) = P\{Y \leq y\} = P\{X^2 \leq y\} = P\{-\sqrt{y} \leq X \leq \sqrt{y}\}$$

$$= \int_{-\sqrt{y}}^{\sqrt{y}} f_X(x)dx = F_X(\sqrt{y}) - F_X(-\sqrt{y})$$

$$f_Y(y) = F_Y'(y) = f_X(\sqrt{y})(\sqrt{y})' - f_X(-\sqrt{y})(-\sqrt{y})'$$

$$= \frac{1}{2\sqrt{y}}\left[f_X(\sqrt{y}) + f_X(-\sqrt{y})\right]$$

Therefore, the probability density of Y is

$$f_Y(y) = \begin{cases} \frac{1}{2\sqrt{y}}\left[f_X(\sqrt{y}) + f_X(-\sqrt{y})\right] & y > 0 \\ 0 & y \leq 0 \end{cases}$$

If X ~ N (0, 1), and the probability density is

$$f_X(x) = \frac{1}{\sqrt{2\pi}} e^{-\frac{x^2}{2}} \qquad -\infty < x < \infty$$

then the probability density of $Y = X^2$ is

$$f_Y(y) = \frac{1}{2\sqrt{y}}\left[f_X(\sqrt{y}) + f_X(-\sqrt{y})\right] = \frac{1}{\sqrt{2\pi}} y^{-\frac{1}{2}} e^{-\frac{y}{2}}$$

That is,
$$f_Y(y) = \begin{cases} \frac{1}{\sqrt{2\pi}} y^{-\frac{1}{2}} e^{-\frac{y}{2}} & y > 0 \\ 0 & y \leq 0 \end{cases}$$

At this time, Y is said to obey the χ^2 distribution with a degree of freedom of 1. That is, $Y \sim \chi^2(1)$.

Theorem

Let the random variable X have a probability density $f_X(x)$, $-\infty < x < \infty$. Let the function $y = g(x)$ be strictly monotonous (increasing or decreasing), differentiable and $g'(x) > 0$ (or $g'(x) < 0$). Then, Y = g(X) is a continuous random variable. According to the differentiation rule of definite integrals, the probability density of Y is

$$f_Y(y) = \begin{cases} f_X(x)\left|\dfrac{dx}{dy}\right| & \alpha < y < \beta \\ 0 & elsewhere \end{cases} \quad \text{or} \quad f_Y(y) = \begin{cases} f_X\left[h(y)\right]\left|h'(y)\right| & \alpha < y < \beta \\ 0 & elsewhere \end{cases}$$

where the inverse function $x = h(y)$, $\left|\dfrac{dx}{dy}\right| = \left|h'(y)\right|$. (The absolute value is taken to satisfy $f_Y(y) \geq 0$.)

Proof. According to the substitution method of the definite integral, we have

$$F_Y(y) = P\{Y \leq y\} = \int\limits_{g(x)\leq y} f_X(x)dx = \int\limits_{g(x)\leq y} f_X(x)\frac{dy}{dy}dx = \int_{-\infty}^{y} f_X(x)\frac{dx}{dy}dy$$

Hence $$f_Y(y) = \frac{dF_Y(y)}{dy} = f_X(x)\left|\frac{dx}{dy}\right|$$ □

Note:

Not all functions have inverse functions $x = g^{-1}(y) = h(y)$. For example, the function $y = x^{2k}$ (k = 1, 2, …) does not have an inverse function. This theorem fails for Y = g(X) without an inverse function. It can only be done according to the method in Example 3, that is, according to the rule of differentiation of definite integrals.

Example 4. Let the random variable X ~ N (μ, σ^2). Prove that the linear function

Y = $aX + b$ $(a \neq 0)$ of X also obeys the normal distribution.

Proof. The probability density of X is

$$f_X(x) = \frac{1}{\sigma\sqrt{2\pi}} e^{-\frac{1}{2}\left(\frac{x-\mu}{\sigma}\right)^2} \qquad -\infty < x < \infty$$

Since $\quad y = g(x) = ax + b \qquad x = h(y) = \dfrac{y-b}{a}$

$$\frac{dy}{dx} = a$$

Cross-multiplying gives

$$\frac{dx}{dy} = \frac{1}{a}$$

So, the probability density of Y = $aX + b$ is

$$f_Y(y) = f_X(x)\left|\frac{dx}{dy}\right| = f_X(\frac{y-b}{a})\frac{1}{|a|} = \frac{1}{|a|}\frac{1}{\sigma\sqrt{2\pi}} e^{-\frac{1}{2}\left(\frac{\frac{y-b}{a}-\mu}{\sigma}\right)^2}$$

That is, $\quad f_Y(y) = \dfrac{1}{|a|\sigma\sqrt{2\pi}} e^{-\frac{[y-(b+a\mu)]^2}{2(a\sigma)^2}} \qquad -\infty < y < \infty$

Therefore, we have

$$Y = aX + b \sim N(a\mu + b, (a\sigma)^2)$$

In particular, if $a = \dfrac{1}{\sigma}$, $b = -\dfrac{\mu}{\sigma}$, then

$$Y = \frac{X-\mu}{\sigma} \sim N(0, 1)$$

Example 5. Let the voltage Y = AsinX, where A is a known positive constant, and the phase angle X is a random variable that has a uniform distribution on the interval $(-\frac{\pi}{2}, \frac{\pi}{2})$. Find the probability density of voltage Y.

Solution. The probability density of X is

$$f_X(x) = \begin{cases} \dfrac{1}{\pi} & -\dfrac{\pi}{2} < x < \dfrac{\pi}{2} \\ 0 & elsewhere \end{cases}$$

Because on $(-\frac{\pi}{2}, \frac{\pi}{2})$, $y = g(x) = \mathrm{A}\sin x$ is a monotonous function so

$x = h(y) = \arcsin \frac{y}{\mathrm{A}}$.

$$\frac{dy}{dx} = \mathrm{A}\cos x > 0$$

Cross-multiplying gives

$$\frac{dx}{dy} = \frac{1}{\mathrm{A}\cos x} = \frac{1}{\mathrm{A}\sqrt{1-\sin^2 x}} = \frac{1}{\mathrm{A}\sqrt{1-\left(\frac{y}{\mathrm{A}}\right)^2}} = \frac{1}{\sqrt{\mathrm{A}^2 - y^2}}$$

So, the probability density of Y = AsinX is

$$f_Y(y) = f_X(x)\left|\frac{dx}{dy}\right| = f_X(\arcsin\frac{y}{\mathrm{A}})\frac{1}{\sqrt{\mathrm{A}^2 - y^2}} = \frac{1}{\pi}\cdot\frac{1}{\sqrt{\mathrm{A}^2 - y^2}}$$

That is,

$$f_Y(y) = \begin{cases} \dfrac{1}{\pi\sqrt{\mathrm{A}^2 - y^2}} & -\mathrm{A} < y < \mathrm{A} \\ 0 & elsewhere \end{cases}$$

where $x \in (-\frac{\pi}{2}, \frac{\pi}{2})$, $\frac{y}{\mathrm{A}} = \sin x \in (-1, 1) \quad \Rightarrow \quad -\mathrm{A} < y < \mathrm{A}$ □

If, in this example, X has a uniform distribution on (0, π), then $y = g(x)$ = Asin$x$ is not a monotonous function on (0, π). The above theorem is invalid, and we can only analyse concretely to solve it. Next, let us solve one problem like this.

Example 6. Let the voltage Y = AsinX, where A is a known positive constant, and the phase angle X is a random variable that has a uniform distribution on the interval $(0, \pi)$. Find the probability density of voltage Y.

Solution. When $x \in (-\frac{\pi}{2}, \frac{\pi}{2})$ or $x \in (0, \pi)$, X has the same probability density.

$$f_X(x) = \begin{cases} \frac{1}{\pi} & -\frac{\pi}{2} < x < \frac{\pi}{2} \quad or \quad 0 < x < \pi \\ 0 & elsewhere \end{cases}$$

We know from Example 5 that on $x \in (-\frac{\pi}{2}, \frac{\pi}{2})$, the function $y = \text{A}\sin x$ is a monotonous function, and its probability density is

$$f_Y(y) = \begin{cases} \frac{1}{\pi\sqrt{\text{A}^2 - y^2}} & -\text{A} < y < \text{A} \\ 0 & elsewhere \end{cases} \qquad (f_Y(y) \text{ is an even function.})$$

On $x \in (0, \pi)$, the function $y = \text{A}\sin x$ is not monotonous because $\frac{dy}{dx} = \text{A}\cos x < 0$ (or > 0). Hence, the above theorem is invalid, but it still has a probability density, which is

$$\begin{cases} f_Y^*(y) > 0 & 0 < y < \text{A} \\ f_Y^*(y) = 0 & elsewhere \end{cases}$$

where $x \in (0, \pi)$, $\frac{y}{\text{A}} = \sin x \in (0, 1) \Rightarrow 0 < y < \text{A}$.

According to the properties of probability density, we have

$$1 = \int_{-\text{A}}^{\text{A}} f_Y(y)dy = 2\int_0^{\text{A}} f_Y(y)dy \qquad (f_Y(y) \text{ is an even function.})$$

$$1 = \int_0^{\text{A}} f_Y^*(y)dy$$

That is, $$2\int_0^A f_Y(y)dy = 1 = \int_0^A f_Y^*(y)dy$$

Then $$f_Y^*(y) = 2f_Y(y) = \frac{2}{\pi\sqrt{A^2 - y^2}}$$

So on $x \in (0, \pi)$, the probability density of non-monotonous function Y = AsinX is

$$f_Y^*(y) = \begin{cases} \dfrac{2}{\pi\sqrt{A^2 - y^2}} & 0 < y < A \\ 0 & elsewhere \end{cases}$$

Exercise 3.5

1. Let the random variable X have the following probability distribution.

X	-2	-1	0	1
p_k	0.1	0.4	0.2	0.3

Find the probability distribution of Y = |X + 1|.

2. The random variable X has the following probability density

$$f_X(x) = \begin{cases} 1 & 0 < x < 1 \\ 0 & elsewhere \end{cases}$$

Find the probability density of random variable Y = e^X.

3. Random variable X obeys a uniform distribution on $(-\frac{\pi}{2}, \frac{\pi}{2})$. Find the probability density of the following functions

a. Y = tanX.

b. Y = cosX.

3.6. Examples

Example 1. Conduct repeated independent experiments, that is, experiments of sampling with replacement or equivalent. Let the probability of success in each trial be p, and the probability of failure be $q = 1 - p$ $(0 < p < 1)$.

(1) Conduct the experiments until one success occurs. Let X represent the number of trials required. Find the distribution of X (at this time, we say that X has the geometric distribution with p as the parameter).

(2) Continue the experiments until k successes occur. Let Y represent the number of trials required. Find the distribution of Y (in this case, we say that Y has the Pascal distribution with k and p as parameters or negative binomial distribution).

(3) A basketball player shoots 43% from the field. Let X represent the cumulative number of field goal attempts he first successfully shot. Write down the distribution of X. And calculate the probability that X takes an even number.

Solution. (1) Conduct the experiments until one success occurs. Find the probability distribution (geometric distribution) for the number of trials X = n.

Because

the event $\{X = n\}$ occurs

= $\{n - 1$ failures$\}$ and $\{$success at the last time$\}$ occur simultaneously

so

$$P\{X = n\} = \underbrace{q \cdot q \cdots q}_{n-1\ \ pieces} \cdot p = pq^{n-1} = p(1-p)^{n-1}$$

That is,

$$P\{X = n\} = p(1-p)^{n-1} \qquad n = 1, 2, \ldots$$

(2) First, let us read one specific example: Conduct the experiments until $k = 2$ successes occur. The number of trials is Y = 3. Find the probability.

Let $a = \{\text{success}\}$ $\quad a' = \{\text{failure}\}$

$$\{Y = 3\} \text{ occurs} = \begin{cases} aa'a & occurs \\ & or \\ a'aa & occurs \end{cases}$$

Therefore $\quad P\{Y = 3\} = p^2q + p^2q = 2p^2q = \binom{3-1}{2-1} p^2(1-p)^{3-2}$

According to this example, we can get by Induction: Carry out the experiments until *k* successes occur. The probability distribution (Pascal distribution or negative binomial distribution) of the number of trials Y = *n* is

$$P\{Y = n\} = \binom{n-1}{k-1} p^k (1-p)^{n-k} \qquad k \le n = 1, 2, \ldots$$

(3) X follows a geometric distribution. The distribution is

$$P\{X = n\} = p(1-p)^{n-1} = 0.43(0.57)^{n-1} \qquad n = 1, 2, \ldots$$

Since

$$\{\text{X takes an even number}\} \text{ occurs} = \begin{cases} \{X = 2\} & occurs \\ & or \\ \{X = 4\} & occurs \\ & or \\ & \vdots \end{cases}$$

so

$$P\{\text{X takes an even number}\} = \sum_{k=1}^{\infty} P\{X = 2k\} = \sum_{k=1}^{\infty} 0.43(0.57)^{2k-1} = \frac{0.43 \cdot 0.57}{1-0.57^2} = \frac{57}{157}$$

(sum of a geometric sequence)

That is, the probability that X takes an even number X is

$$P\{\text{X takes an even number}\} = \frac{57}{157}$$

Example 2*. (Banach's matchbox problem) A mathematician who smokes always carries two matchboxes with him: one in his left pocket and one in his right pocket. Each time he needs a match, he is equally likely to take it from either pocket and take out one match. At the beginning, both matchboxes contain N matches. When he discovers that the box picked is empty for the first time, what is the probability that there are exactly k matches in the other box? k = 0, 1, 2,…, N.

Solution. The probability of randomly picking a box from two pockets is $p=\frac{1}{2}$.

Let A = {It is found that the matchbox in the right pocket is empty, and the matchbox in the left pocket contains k matches.}

Because

the matchbox in the right pocket is found to be empty

= the matchbox has been drawn N + 1 times

the matchbox in the left pocket contains k matches

= the matchbox has been drawn N – k times

so the total number of draws is

$$n = \mathrm{N} + 1 + \mathrm{N} - \mathrm{k} = 2\mathrm{N} - \mathrm{k} + 1$$

until he found the matchbox in the right pocket empty, $r = \mathrm{N} + 1$.

According to the negative binomial distribution formula, the probability of A is

$$\begin{aligned}\mathrm{P(A)} &= \binom{n-1}{r-1} p^r (1-p)^{n-r} \\ &= \binom{2\mathrm{N}-\mathrm{k}+1-1}{\mathrm{N}+1-1}\left(\frac{1}{2}\right)^{\mathrm{N+1}}\left(\frac{1}{2}\right)^{2\mathrm{N-k+1-(N+1)}} \\ &= \binom{2\mathrm{N}-\mathrm{k}}{\mathrm{N}}\left(\frac{1}{2}\right)^{2\mathrm{N-k+1}}\end{aligned}$$

Let

C occurs

= {The first time he finds a box empty and another box with k matches} occurs

$$=\begin{cases} \textit{Left matchbox is empty and right matchbox has } \mathrm{k} \textit{ matches.} \\ \textit{or} \\ \textit{Right matchbox is empty and left matchbox has } \mathrm{k} \textit{ matches.} \end{cases}$$

Because of symmetry, the probability of C is

$$P(C) = 2P(A) = \binom{2N-k}{N}\left(\frac{1}{2}\right)^{2N-k}$$

Example 3*. There are N balls in a box, including M special balls, $r = N - M$ ordinary balls. Each time, one ball is randomly drawn from the box using sampling without replacement. Let Y = n be the number of times required until k special balls have been drawn. Then, the distribution of Y is called a negative hypergeometric distribution. Next, we will derive its probability distribution formula.

Because

the event {Y = n} occurs = {draw (n – 1) times and get (k – 1) special balls} and {draw 1 time at the end and get 1 special ball} occur simultaneously

so the probability formula of negative hypergeometric distribution is

$$P\{Y = n\} = \frac{\binom{M}{k-1}\binom{N-M}{n-k}}{\binom{N}{n-1}} \cdot \frac{\binom{M-(k-1)}{1}}{\binom{N-(n-1)}{1}}$$

where in combination $n - k = (n - 1) - (k - 1)$

In particular, when M and N are both very large relative to k and n, sampling without replacement can be approximately treated as sampling with replacement. At this time, the negative hypergeometric distribution of X is approximately equal to the negative binomial distribution.

That is,
$$P\{Y=n\}=\frac{\binom{M}{k-1}\binom{N-M}{n-k}}{\binom{N}{n-1}}\cdot\frac{\binom{M-(k-1)}{1}}{\binom{N-(n-1)}{1}}\approx\binom{n-1}{k-1}p^k(1-p)^{n-k}$$

where $p=\dfrac{M}{N}$. M and N are both very large relative to k and n.

Suppose there are 10 special balls and 15 ordinary balls in the box. One ball is randomly drawn from the box each time using sampling without replacement. Find the probability that it takes 8 draws until 5 special balls have been drawn.

Solution. According to the formula, we get

$$P\{Y=8\}=\frac{\binom{10}{5-1}\binom{25-10}{8-5}}{\binom{25}{8-1}}\cdot\frac{\binom{10-(5-1)}{1}}{\binom{25-(8-1)}{1}}=\frac{637}{9614}\approx 0.0663$$

Example 4. There is a busy bus station with many buses passing through every day. The probability that each bus will have an accident at some time during the day is 0.00015. 900 buses pass through at that time on a particular day. What is the probability that the number of accidents will be at least 2?

Solution. Let in a day, each bus be X_i.

$$X_i=\begin{cases}1 & \textit{has accident} \quad p=0.00015\\ 0 & \textit{no accident} \quad q=1-p=0.99985\end{cases} \qquad i=1,2,\ldots,900$$

In a day, 900 buses have accidents in total $X=k$ times.

$$X = \sum_{i=1}^{900} X_i \sim B\ (900,\ 0.00015) \qquad \text{follows a binomial distribution.}$$

Because $\qquad 20 < n = 900, \qquad 0.05 > p = 0.00015$

so we can use the Poisson theorem to calculate. Here $\lambda = np = 0.135$. The formula is

$$P\{X = k\} = \frac{\lambda^k e^{-\lambda}}{k!}$$

The probability that the number of accidents will be at least 2 is

$$P\{X \geq 2\} = 1 - P\{0 \leq X < 2\} = 1 - (P\{X = 0\} + P\{X = 1\})$$

$$= 1 - (\frac{0.135^0 e^{-0.135}}{0!} + \frac{0.135^1 e^{-0.135}}{1!}) \approx 0.00833$$

That is, $\qquad P\{X \geq 2\} \approx 0.00833$

Example 5. There are 5 elevators in one building. An investigation shows that the probability of each elevator being used at any given moment t is 0.2. At the same moment

(1) what is the probability that exactly 2 elevators are being used?

(2) what is the probability that at least 3 elevators are being used?

Solution. Let each elevator be X_i.

$$X_i = \begin{cases} 1 & in \quad use \qquad p = 0.2 \\ 0 & not \quad in \quad use \qquad q = 1 - p = 0.8 \end{cases} \qquad i = 1, 2, \ldots, 5$$

At the same moment, there are $X = k$ elevators in use.

$$X = \sum_{i=1}^{5} X_i \sim B\ (5,\ 0.2) \qquad \text{follows a binomial distribution.}$$

The formula for binomial distribution is

$$P\{X = k\} = \binom{n}{k} p^k (1-p)^{n-k}$$

(1) The probability that exactly 2 elevators are used is

$$P\{X=2\}=\binom{5}{2}0.2^2(1-0.2)^{5-2}=0.2048$$

That is, $P\{X=2\}=0.2048$

(2) The probability that at least 3 elevators are used is

$$P\{X\geq 3\}=P\{X=3\}+P\{X=4\}+P\{X=5\}$$

$$=\binom{5}{3}0.2^3(1-0.2)^{5-3}+\binom{5}{4}0.2^4(1-0.2)^{5-4}+\binom{5}{5}0.2^5(1-0.2)^{5-5}$$

$$=0.05792$$

That is, $P\{X\geq 3\}=0.05792$

Example 6. A jury consists of 9 jurors for a trial. To convict the defendant, at least 6 jurors must agree he is guilty. Let the jurors' judgements be independent of each other, and the probability that each juror makes the correct judgement is p. If it is known that the probability that the defendant is indeed guilty is a, what is the probability that this jury will make a correct verdict?

Solution. Let each juror be X_i.

$$X_i=\begin{cases}1 & \textit{makes correct judgement} & \textit{probability}=p\\ 0 & \textit{makes incorrect judgement} & \textit{probability}=1-p\end{cases}\qquad i=1,2,\ldots,9$$

Among the 9 jurors, X people made the correct judgment. That is,

$$X=\sum_{i=1}^{9}X_i\sim B\,(9,p)\qquad \text{follows a binomial distribution.}$$

The formula of binomial distribution is

$$P\{X=k\}=\binom{n}{k}p^k(1-p)^{n-k}$$

Let A = {The defendant is guilty, and the jury reaches the correct verdict}. Then, the probability of A is

$$P(A) = P\{X \geq 6\} = \sum_{k=6}^{9} \binom{9}{k} p^k (1-p)^{9-k}$$

Let B = {The defendant is not guilty, and the jury reaches the correct verdict}. Then, the probability of B is

$$P(B) = P\{X \geq 4\} = \sum_{k=4}^{9} \binom{9}{k} p^k (1-p)^{9-k}$$

Let

C occurs = {The jury makes a correct verdict} occurs

$$= \begin{cases} \textit{The defendant is guilty} \quad (\textit{probability is } a) \\ \quad \textit{and } A \textit{ occur} \quad \text{simultaneously.} \\ \quad \textit{or} \\ \textit{The defendant is not guilty} \quad (\textit{probability is } 1-a) \\ \quad \textit{and } B \textit{ occur} \quad \text{simultaneously.} \end{cases}$$

Then, the probability of C is

$$P(C) = a\sum_{k=6}^{9} \binom{9}{k} p^k (1-p)^{9-k} + (1-a)\sum_{k=4}^{9} \binom{9}{k} p^k (1-p)^{9-k}$$

Example 7. The lifetime (in hours) of the tubes produced in a factory follows a normal distribution with parameters $\mu = 160$ and σ. What is the maximum allowable σ if $P\{120 < X < 200\} \geq 0.80$ is required?

Solution. Because it is a normal distribution so we have the formula

$$P\{x_1 \leq X \leq x_2\} = \Phi\left(\frac{x_2-\mu}{\sigma}\right) - \Phi\left(\frac{x_1-\mu}{\sigma}\right) \qquad \Phi(-x) = 1 - \Phi(x)$$

so
$$P\{120 < X < 200\} = \Phi\left(\frac{200-160}{\sigma}\right) - \Phi\left(\frac{120-160}{\sigma}\right)$$

$$= \Phi\left(\frac{40}{\sigma}\right) - \Phi\left(\frac{-40}{\sigma}\right) = 2\Phi\left(\frac{40}{\sigma}\right) - 1$$

It is required that

$$P\{120 < X < 200\} \geq 0.80$$

$$2\Phi\left(\frac{40}{\sigma}\right) - 1 \geq 0.80$$

Therefore $$\Phi\left(\frac{40}{\sigma}\right)_{\min} = 0.9$$

Using a calculator, we get

$$\frac{40}{\sigma} = 1.28$$

So $$\sigma_{\max} = \frac{40}{1.28} = 31.25$$

Example 8. Let k follow a uniform distribution on (0, 5). Find the probability that the following equation has real roots.

$$4x^2 + 4kx + k + 2 = 0$$

Solution. If the equation has real roots, then the discriminant of the quadratic equation will be greater than or equal to 0. That is,

$$b^2 - 4ac = (4k)^2 - 4 \cdot 4(k + 2) \geq 0$$

$$16(k^2 - k - 2) \geq 0$$

That is, $$k^2 - k - 2 \geq 0 \quad \Rightarrow \quad k(k - 1) \geq 2 \quad \Rightarrow \quad k \geq 2$$

Therefore, when $k \geq 2$, the equation will have real roots.

Because the probability density of k is

$$f_K(k) = \frac{1}{5-0} = \frac{1}{5}$$

That is, $$f_K(k) = \begin{cases} \frac{1}{5} & 0 < k < 5 \\ 0 & \textit{elsewhere} \end{cases}$$

So, the probability that the equation has real roots is

$$P\{k \ge 2\} = \int_2^5 \frac{1}{5} dx = \frac{3}{5}$$

That is, $$P\{k \ge 2\} = \frac{3}{5}$$

Example 9. Let the time X (in minutes) that a customer waits for service at a bank follow an exponential distribution. The probability density is

$$f_X(x) = \begin{cases} \frac{1}{5} e^{-\frac{x}{5}} & x > 0 \\ 0 & elsewhere \end{cases}$$

A customer waits for service at the bank. If it takes more than 10 minutes, he leaves. He goes to the bank five times a month. Let Y represent the number of times in a month that he leaves the bank without getting the service. Write down the distribution of Y and find $P\{Y \ge 1\}$.

Solution. The probability of waiting for service for more than 10 minutes is

$$P\{X \ge 10\} = \int_{10}^{\infty} \frac{1}{5} e^{-\frac{x}{5}} dx = \left[-e^{-\frac{x}{5}} \right]_{10}^{\infty} = e^{-2}$$

So $$Y_i = \begin{cases} 1 & not \quad waiting \quad for \quad service & p = e^{-2} \\ 0 & waiting \quad for \quad service & q = 1 - p = 1 - e^{-2} \end{cases}$$

$$Y = \sum_{i=1}^{5} Y_i \sim B(5, e^{-2})$$ follows a binomial distribution.

The distribution of Y is

$$P\{Y = k\} = \binom{n}{k} p^k (1-p)^{n-k} = \binom{5}{k} e^{-2k} (1-e^{-2})^{5-k}$$

That is, $$P\{Y = k\} = \binom{5}{k} e^{-2k} (1-e^{-2})^{5-k} \qquad k = 0, 1, 2, \ldots, 5$$

Then $$P\{Y \ge 1\} = 1 - P\{Y = 0\} = 1 - (1-e^{-2})^5 \approx 0.517$$

That is, $P\{Y \geq 1\} \approx 0.517$

Example 10. A person was accused of being the father of a newborn child. The expert in the case testified that the number of days during the mother's pregnancy approximately follows a normal distribution, with the parameters $\mu = 270$ days for the average pregnancy and $\sigma^2 = 100$ days for the variance. The testimony given by the defendant indicates that he went abroad 280 days before the birth of the child and did not return until 250 days before the birth. If the defendant is the child's father, find the probability that the child was born within the time period consistent with the defendant's testimony.

Solution. Let X denote the number of days of pregnancy. If the defendant is the father of the child, then

let $A = \{$The child was born within the time period consistent with the defendant's testimony$\}$

$$= \begin{cases} \{X > 280\} & occurs \\ \quad or & \\ \{X < 250\} & occurs \end{cases}$$

Because $X \sim N(270, 100)$ follows a normal distribution.

so $P(A) = P\{X > 280\} + P\{X < 250\} \approx 0.181$

That is, $P(A) \approx 0.181$

Example 11*. Let current $Y = A\sin(\omega X)$. The current Y is a random variable. It is uniformly distributed between 9 amps and 11 amps. The random variable X is time. A, ω are constants. If this current passes through a 2-ohm resistor, the power consumed is $Z = 2Y^2$. Find the probability density of Z.

Solution. Because $Y = A\sin(\omega X)$ so $Z = 2Y^2 = 2[A\sin(\omega X)]^2$

Since the probability density formula of the function Y = g(X) is

$$f_Y(y) = f_X(x)\left|\frac{dx}{dy}\right|$$

so first find

$$\frac{dy}{dx} = A\omega\cos(\omega x)$$

$$\frac{dz}{dx} = 4\omega A^2 \sin(\omega x)\cos(\omega x)$$

Cross-multiplying gives

$$\frac{dx}{dz} = \frac{1}{4\omega A^2 \sin(\omega x)\cos(\omega x)}$$

And $$f_Y(y) = \frac{1}{11-9} = \frac{1}{2}$$

That is, $$f_Y(y) = \begin{cases} \frac{1}{2} & 9 < y < 11 \\ 0 & elsewhere \end{cases}$$

$$f_X(x) = f_Y(y)\left|\frac{dy}{dx}\right| = \frac{1}{2}A\omega\cos(\omega x)$$

$$f_Z(z) = f_X(x)\left|\frac{dx}{dz}\right| = \frac{1}{2}A\omega\cos(\omega x)\cdot\frac{1}{4\omega A^2 \sin(\omega x)\cos(\omega x)}$$

$$= \frac{1}{8}\cdot\frac{1}{A\sin(\omega x)} = \frac{1}{8}\left(\frac{2}{z}\right)^{\frac{1}{2}}$$

Therefore $$f_Z(z) = \frac{1}{8}\left(\frac{2}{z}\right)^{\frac{1}{2}} \qquad 162 < z < 242$$

where $$9 < \sqrt{\frac{z}{2}} < 11 \qquad \Rightarrow \qquad 162 < z < 242$$

Example 12*. Let the probability density of random variable X be

$$f(x)=\begin{cases}\dfrac{2x}{\pi^2} & 0<x<\pi \\ 0 & elsewhere\end{cases}$$

Find the probability density of Y = sinX.

Solution. Because on $x \in (0, \pi)$, Y = sinX is not a monotonous function. The formula

$$f_{\mathrm{Y}}(y)=f_{\mathrm{X}}(x)\left|\frac{dx}{dy}\right|$$

is invalid. Let us analyse concretely to solve.

Because $$1=\int_0^{\pi}\frac{2x}{\pi^2}dx=\int_0^{\frac{\pi}{2}}\frac{2x}{\pi^2}dx+\int_{\frac{\pi}{2}}^{\pi}\frac{2x}{\pi^2}dx$$

then $$\int_{\frac{\pi}{2}}^{\pi}\frac{2x}{\pi^2}dx=1-\int_0^{\frac{\pi}{2}}\frac{2x}{\pi^2}dx=1-\frac{1}{4}=\frac{3}{4}$$

That is, we have

$$1=\mathrm{P}\{0<\mathrm{X}<\pi\}\approx\mathrm{P}\{\frac{\pi}{2}<\mathrm{X}<\pi\}=0.75$$

On $x \in (\frac{\pi}{2}, \pi)$, Y = sinX is a monotonous function.

From $\frac{\pi}{2}<x<\pi \quad \Rightarrow \quad$ Let $x\approx\pi$ then

$$f_{\mathrm{Y}}^{*}(y)=f_{\mathrm{X}}(x)\left|\frac{dx}{dy}\right|=\frac{2x}{\pi^2}\cdot\frac{1}{\sqrt{1-y^2}}\approx\frac{2}{\pi}\cdot\frac{1}{\sqrt{1-y^2}} \qquad \frac{\pi}{2}<x<\pi, \quad 0<y<1$$

Since $$\mathrm{P}\{0<\mathrm{X}<\pi\}\approx\mathrm{P}\{\frac{\pi}{2}<\mathrm{X}<\pi\}$$

so $$f_{\mathrm{Y}}(y)\approx f_{\mathrm{Y}}^{*}(y)$$

That is, the probability density of the non-monotonous function Y = sinX on

$x \in (0, \pi)$ is

$$f_Y(y) \approx \begin{cases} \dfrac{2}{\pi\sqrt{1-y^2}} & 0 < x < \pi, \quad 0 < y < 1 \\ 0 & elsewhere \end{cases}$$

Exercise 3.6

1. We observe whether event A occurs in a series of independent experiments. In each experiment, the probability of event A occurring is 0.4. Let X denote the total number of experiments completed when event A occurs.

a. Find the probability distribution of X.

b. Find the probability that X is an even number.

c. Find the probability that X > 2.

2. Conduct four independent experiments. In each experiment, the probability that outcome A occurs is 0.3. If A does not occur, then B does not occur. If A occurs only once, the probability that B occurs is 0.6. If A occurs at least twice, the probability that B occurs is 1.

a. Find the probability distribution of occurrence of A.

b. Find the probability that B occurs.

c. Given that B occurs, find the probability that A occurs exactly once.

3. A class has 15 boys and 20 girls. Five students are chosen at random to visit a museum. Find the probability that at least two students of each sex are selected.

4. Three cards are drawn one after another at random without replacement from a box containing 10 cards numbered 1, 2, 3,…, 10. Find the probability that the largest number is six.

5. Urn A contains three white balls and three red balls. Urn B contains 5 white balls and 1 red ball. An urn is selected randomly, and three balls are selected randomly without replacement from the urn. If two of the three chosen balls are white, what is the probability that the urn selected was urn A?

6. The random variables X and Y follow binomial distributions B (2, p) and B (3, 2p), respectively. $P\{X \geq 1\} = 0.51$. Find the value of $P\{Y \geq 1\}$.

7. Basketball team A plays basketball against team B. For each round, the probability that team A wins is 0.6, and the probability that team B wins is 0.4. Each round is independent. It is agreed that the team which wins four rounds first will get the championship. Find the probability p_i that team A will get the tournament after i (i = 4, 5, 6, 7) rounds.

8. The random variable X follows a uniform distribution on (-5, 5). Find the probability that the equation

$$x^2 + Xx + 1 = 0$$

has real root(s).

9. Workers in a factory incur accidents at the rate of two accidents per week. Calculate the probability that there will be at most two accidents

- **a.** during 1 week.
- **b.** during 2 weeks.
- **c.** in each of 2 weeks.

10. The random variable X follows a uniform distribution on (1, 4). Make three independent observations of X. Find the probability that at least two values observed are greater than 2.

11. The volume of tomato sauce poured into cans in a production process follows a normal distribution with a standard deviation of 4.2 ml. Among all the cans, 25% contain less than 700 ml.

a. Find the mean volume of tomato sauce in the cans. Give your answer correct to two decimal places.

b. 10 cans are chosen at random. Find the probability that at most two cans contain less than 698 ml.

Chapter 4 Numerical characteristics of random variables

We discussed cumulative distribution functions (sum event functions) in previous chapters. It can be seen that the cumulative distribution function can provide a complete description of the statistical properties of a random variable. However, in some practical situations, it is not necessary to consider the cumulative distribution function of a random variable. We only need to know the numerical characteristics of the random variable.

4.1. Mathematical expectation

Let us start with an example. A man is practising shooting. If the shot hits the region e_2 of the target, as shown in the following diagram, he gets two points. If the shot hits the region e_1, he gets one point. If the shot misses the target, i.e. hits the region e_0, he gets zero points. Let the random variable X be the "number of points he gets for each shot", that is,

$$\text{Number of points he gets for each shot } \mathrm{X}=k=\begin{cases} 0 & (0 \ \ represents \ \ e_0) \qquad p_0 \\ 1 & (1 \ \ represents \ \ e_1) \qquad p_1 \\ 2 & (2 \ \ represents \ \ e_2) \qquad p_2 \end{cases}$$

Probability distribution of X $\quad \mathrm{P}\{\mathrm{X}=k\}=p_k \quad k=0,1,2$

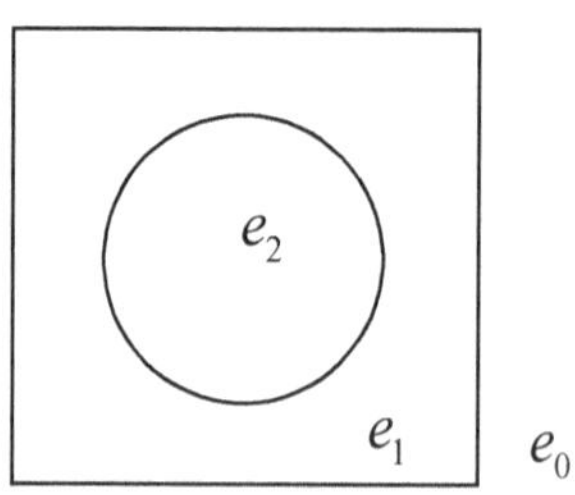

He shoots n times, among which he gets 0 points for f_0 times, 1 point for f_1 times, and 2 points for f_2 times. $f_0+f_1+f_2=n$. The sum of the points he gets in n-shots is $0\cdot f_0+1\cdot f_1+2\cdot f_2$. Therefore, the arithmetic average of each shot is $\overline{x}$.

arithmetic average $\overline{x}=\dfrac{0\times f_0+1\times f_1+2\times f_2}{n}=0\times\dfrac{f_0}{n}+1\times\dfrac{f_1}{n}+2\times\dfrac{f_2}{n}=\sum_{k=0}^{2}k\times\dfrac{f_k}{n}$

Here $f_n\{X=k\}=\dfrac{f_k}{n}$ is the relative frequency of the event $\{X=k\}$. When the number of shots n differs, the relative frequency $\dfrac{f_k}{n}$ fluctuates randomly between 0 and 1. If $n\neq\infty$, that is, the number of shots n is only equivalent to a partial number of shots, then $\dfrac{f_k}{n}\neq$ constant. Therefore, we cannot be sure that the arithmetic average $\overline{x}$ is the exact average of all possible values $\begin{cases}0\\1\\2\end{cases}$. Only when the total number of shots $n=\infty$, $\lim\limits_{n\to\infty}\dfrac{f_k}{n}=p_k=$ constant. At this time, we can be sure that the average of all possible values $\begin{cases}0\\1\\2\end{cases}$ is the exact average. That is,

$$\lim_{n\to\infty}\overline{x}=\lim_{n\to\infty}(0\times\frac{f_0}{n}+1\times\frac{f_1}{n}+2\times\frac{f_2}{n})=\lim_{n\to\infty}\sum_{k=0}^{2}k\times\frac{f_k}{n}$$

$$=0\times\lim_{n\to\infty}\frac{f_0}{n}+1\times\lim_{n\to\infty}\frac{f_1}{n}+2\times\lim_{n\to\infty}\frac{f_2}{n}=\sum_{k=0}^{2}k\times\lim_{n\to\infty}\frac{f_k}{n}$$

$$=0\times p_0+1\times p_1+2\times p_2=\sum_{k=0}^{2}k\times p_k=\sum_{k=0}^{2}k\times P\{X=k\}$$

That is, $$\lim_{n\to\infty}\overline{x}=\sum_{k=0}^{2}k\mathrm{P}\{X=k\}=\sum_{k=0}^{2}kp_k$$

Definition of mathematical expectation

(1) Let the probability distribution of discrete random variable X be

$$\mathrm{P}\{\mathrm{X}=x_k\}=p_k \qquad k=1,2,\ldots$$

If the series

$$\sum_{k=1}^{\infty} x_k p_k$$

converges absolutely (that is, it is a finite number), then according to the definition of average, the average of the range of X is

$$E(\mathrm{X})=\frac{\sum_{k=1}^{\infty} x_k p_k}{\sum_{k=1}^{\infty} p_k}=\sum_{k=1}^{\infty} x_k p_k$$

E(X) is called the mathematical expectation or population mean (or simply expectation or mean) of X.

In particular, if Y is a function of the random variable X, that is, Y = g(X), then we also have

$$E(\mathrm{Y})=E[g(\mathrm{X})]=\frac{\sum_{k=1}^{\infty} g(x_k) p_k}{\sum_{k=1}^{\infty} p_k}=\sum_{k=1}^{\infty} g(x_k) p_k$$

where $\sum_{k=1}^{\infty} p_k=\sum_{k=1}^{\infty} \mathrm{P}\{\mathrm{X}=x_k\}=1$.

(2) Similarly, let the probability density of a continuous random variable X be $f(x)$. If the integral

$$\int_{-\infty}^{\infty} x f(x) dx$$

converges absolutely (a finite number), then we have

$$E(\mathrm{X})=\sum_{k=1}^{\infty}x_k p_k=\sum_{k=1}^{\infty}x_k f(x_k)\lim_{\Delta x_k\to 0}\Delta x_k=\int_{-\infty}^{\infty}xf(x)dx$$

or according to the definition of average, we have

$$E(\mathrm{X})=\frac{\int_{-\infty}^{\infty}xf(x)dx}{\int_{-\infty}^{\infty}f(x)dx}=\int_{-\infty}^{\infty}xf(x)dx$$

$E(\mathrm{X})$ is called the mathematical expectation or population mean (or simply expectation or mean) of X.

In particular, if Y is a function of the random variable X, that is, Y = g(X) (g is a continuous function), then we also have

$$E(\mathrm{Y})=E\left[g(\mathrm{X})\right]=\frac{\int_{-\infty}^{\infty}g(x)f(x)dx}{\int_{-\infty}^{\infty}f(x)dx}=\int_{-\infty}^{\infty}g(x)f(x)dx$$

where $\int_{-\infty}^{\infty}f(x)dx=1$.

Combining the above results, regardless of whether X is discrete or continuous, we have

$$\lim_{n\to\infty}\bar{x}=E(\mathrm{X})$$

When n is relatively large, we have

$$\bar{x}\approx E(\mathrm{X}) \qquad \text{or} \qquad \bar{x}\approx\lim_{n\to\infty}\bar{x}$$

(1) The mathematical expectation of discrete random variables X and Y = g(X) is

$$E(\mathrm{X})=\sum_{k=1}^{\infty}x_k p_k$$

$$E(\mathrm{Y}) = \sum_{k=1}^{\infty} g(x_k) p_k$$

(2) The mathematical expectation of continuous random variables X and Y = g(X) is

$$E(\mathrm{X}) = \int_{-\infty}^{\infty} x f(x) dx$$

$$E(\mathrm{Y}) = \int_{-\infty}^{\infty} g(x) f(x) dx$$

(3) The value of mathematical expectation is denoted as μ. That is,

$$E(\mathrm{X}) = \mu$$

Some important properties of mathematical expectation

(1) If c is a constant, then $E(\mathrm{c}) = \mathrm{c}$.

(2) If X is a random variable and c is a constant, then $E(\mathrm{cX}) = \mathrm{c}E(\mathrm{X})$.

(3) If $\mathrm{X}_1, \mathrm{X}_2, \ldots, \mathrm{X}_n$ are random variables, then

$$E(\sum_{i=1}^{n} \mathrm{X}_i) = E(\mathrm{X}_1) + E(\mathrm{X}_2) + \ldots + E(\mathrm{X}_n)$$

(4) If X and Y are mutually independent random variables, then we have

$$E(\mathrm{XY}) = E(\mathrm{X})E(\mathrm{Y})$$

This property can be extended to the product of any finite number of mutually independent random variables.

(Proof of properties (1) – (4) is omitted.)

Example 1. According to the timetable, there is only one bus arriving at the bus station at 7 : 00 – 8 : 00 every day, but the time of arrival at the bus station is random, and the rule is

arriving time	7 : 10	7 : 30	7 : 50
probability p_k	$\frac{1}{7}$	$\frac{4}{7}$	$\frac{2}{7}$

A passenger arrives at the bus station at 7 : 00. Find the mathematical expectation of the time he will wait for the bus to arrive.

Solution. Let the time the passenger waits for the bus to arrive be X minutes. Because

$$X = \begin{cases} 10 & p_1 = \frac{1}{7} \\ 30 & p_2 = \frac{4}{7} \\ 50 & p_3 = \frac{2}{7} \end{cases}$$

Therefore, the mathematical expectation of waiting time is

$$E(X) = \sum_{k=1}^{\infty} x_k p_k = 10 \times \frac{1}{7} + 30 \times \frac{4}{7} + 50 \times \frac{2}{7} = \frac{230}{7} \approx 32.9 \quad \text{minutes}$$

Example 2. Two people, A and B, perform target shooting. The scores obtained are recorded as X_1 and X_2, respectively. Their distributions are respectively

X_1	0	1	2
p_k	0	0.3	0.7

X_2	0	1	2
p_k	0.5	0.3	0.2

Evaluate their performance.

Solution. Let us calculate the mathematical expectation of X_1.

$$E(X_1) = \sum_{k=1}^{\infty} x_k p_k = 0 \times 0 + 1 \times 0.3 + 2 \times 0.7 = 1.7$$

This means that if A makes a relatively large number of n shots, the arithmetic average of the scores obtained is close to 1.7. That is,

$$\bar{x} \approx E(X_1) = 1.7$$

The mathematical expectation of the score obtained by B is

$$E(X_2) = 0 \times 0.5 + 1 \times 0.3 + 2 \times 0.2 = 0.7$$

Obviously, B's performance is far worse than A's.

Example 3. Let the wind speed X obey a uniform distribution on (0, a). That is, it has a probability density

$$f(x)=\begin{cases}\dfrac{1}{a} & 0<x<a\\ 0 & elsewhere\end{cases}$$

Let the positive pressure Y on the aircraft wing be a function of X. That is, $Y = kX^2$ ($k > 0$, k is a constant). Find the mathematical expectation of Y.

Solution. From the formula, we have

$$E(Y)=\int_{-\infty}^{\infty} g(x)f(x)dx=\int_{0}^{a} kx^2\frac{1}{a}dx=\frac{1}{3}ka^2$$

That is, $$E(Y)=\frac{1}{3}ka^2$$

4.2. Variance

Let us start with an example. A shop has a stock of LED light bulbs in its warehouse. We know that the lifetime of the light bulb is X, and the average lifetime is $E(X) = 13000$ hours. But this alone does not tell us how good or bad the light bulbs are. In fact, at least 80% of these bulbs probably last between 11000 and 15000 hours, meaning that

$$P\{11000 \le X \le 15000\} \ge 80\% = 0.8$$

Or 50% of them may be of high quality and have a lifespan between 13000 and 18000 hours. The remaining 50% are of poor quality and have a lifespan between 8000 and 13000 hours. That is,

$$P\{13000 \le X \le 18000\} = 50\% = 0.5$$

$$P\{8000 \le X \le 13000\} = 50\% = 0.5$$

To evaluate whether the quality of the light bulbs is good, we need to investigate the deviation of the life of light bulbs X from the average value $E(X)$. If the deviation is small, it means that the quality of this batch of light bulbs is stable. And we say the light bulbs are of good quality. So, how do we determine the deviation?

It is easy to observe that

$$E\{|X - E(X)|\}$$

can determine the deviation of a random variable X from its mean $E(X)$. However, this expression contains absolute values and is inconvenient to calculate. For calculation convenience, we use

$$E\{[X - E(X)]^2\}$$

to determine the deviation of a random variable X from its mean $E(X)$.

Definition of variance

Let X be a random variable. If $E\{[X - E(X)]^2\}$ exists, then we say that $E\{[X - E(X)]^2\}$ is the population variance of X (or simply variance) and denote it as Var(X). That is,

$$\text{Var}(X) = E\{[X - E(X)]^2\}$$

In practice, we often need to use the standard deviation, represented by $\sigma(X) = \sigma$.

Standard deviation is the square root of the variance.

$$\sigma = \sqrt{\text{Var}(X)} \qquad \text{so} \qquad \text{Var}(X) = \sigma^2$$

According to the definition of variance, the variance of a random variable X represents the deviation of the values of X from its mathematical expectation $E(X)$. If X takes values from a relatively small interval, then Var(X) will be relatively small.

On the contrary, if the range of X values is relatively large, then Var(X) will be relatively large. Therefore, Var(X) is a measure of the dispersion of X values.

According to the definition of variance, for discrete random variable X

since $$\mathrm{X} = \begin{cases} x_1 & p_1 \\ x_2 & p_2 \\ \vdots & \vdots \end{cases} \qquad \mathrm{X} - E(\mathrm{X}) = \begin{cases} x_1 - E(\mathrm{X}) & p_1 \\ x_2 - E(\mathrm{X}) & p_2 \\ \vdots & \vdots \end{cases}$$

therefore $$\mathrm{Var(X)} = E\{[\mathrm{X} - E(\mathrm{X})]^2\} = \sum_{k=1}^{\infty}[x_k - E(\mathrm{X})]^2 p_k$$

where $\mathrm{P}\{\mathrm{X} = x_k\} = p_k$ is the probability distribution of X. $k = 1, 2, \ldots$

Similarly, for continuous random variable X

$$\mathrm{Var(X)} = E\{[\mathrm{X} - E(\mathrm{X})]^2\} = \int_{-\infty}^{\infty} [x - E(\mathrm{X})]^2 f(x)dx$$

where $f(x)$ is the probability density of X.

We can also use the following formula to calculate the variance of the random variable X.

$$\mathrm{Var(X)} = E(\mathrm{X}^2) - [E(\mathrm{X})]^2$$

Proof. According to the mathematical expectation properties (1), (2) and (3), we have

$$\begin{aligned} \mathrm{Var(X)} &= E\{[\mathrm{X} - E(\mathrm{X})]^2\} \\ &= E\{\mathrm{X}^2 - 2\mathrm{X}E(\mathrm{X}) + [E(\mathrm{X})]^2\} \\ &= E(\mathrm{X}^2) - 2E(\mathrm{X})E(\mathrm{X}) + [E(\mathrm{X})]^2 \\ &= E(\mathrm{X}^2) - [E(\mathrm{X})]^2 \end{aligned}$$

where $E(\mathrm{X}^2) = \sum_{k=1}^{\infty} x_k^2 p_k$ or $E(\mathrm{X}^2) = \int_{-\infty}^{\infty} x^2 f(x)dx$.

Combine the above results.

Variance $\quad \mathrm{Var(X)} = E\{[\mathrm{X} - E(\mathrm{X})]^2\} = E(\mathrm{X}^2) - [E(\mathrm{X})]^2$

standard deviation $\quad \sigma = \sqrt{\mathrm{Var(X)}} \quad$ so $\quad \mathrm{Var(X)} = \sigma^2$

Variance of discrete random variable X

$$\mathrm{Var(X)} = \sum_{k=1}^{\infty}[x_k - E(\mathrm{X})]^2 p_k$$

Variance of continuous random variable X

$$\mathrm{Var(X)} = \int_{-\infty}^{\infty}[x - E(\mathrm{X})]^2 f(x)dx$$

Important properties of variance

(1) If c is a constant, then $\mathrm{Var(c)} = 0$.

(2) If X is a random variable and c is a constant, then $\mathrm{Var(cX)} = \mathrm{c^2Var(X)}$.

(3) If $\mathrm{X}_1, \mathrm{X}_2, \ldots, \mathrm{X}_n$ are mutually independent random variables, then

$$\mathrm{Var}(\sum_{i=1}^{n}\mathrm{X}_i) = \mathrm{Var(X_1)} + \mathrm{Var(X_2)} + \ldots + \mathrm{Var}(\mathrm{X}_n)$$

(Proof of (1) – (3) is omitted.)

4.3. Mathematical expectations and variances of some important random variables

1. (0 – 1) distribution

Let the random variable X follow a (0 – 1) distribution.

$$\mathrm{X} = \begin{cases} 0 \\ 1 \end{cases}$$

Its probability distribution is

$$\mathrm{P}\{\mathrm{X} = 0\} = 1 - p, \quad \mathrm{P}\{\mathrm{X} = 1\} = p$$

So $$E(X) = \sum_{k=1}^{\infty} x_k p_k = 0 \times (1-p) + 1 \times p = p$$

$$E(X^2) = \sum_{k=1}^{\infty} x_k^2 p_k = 0^2 \times (1-p) + 1^2 \times p = p$$

Therefore $\mathrm{Var}(X) = E(X^2) - [E(X)]^2 = p - p^2 = p(1-p)$

That is, $E(X) = p,\qquad \mathrm{Var}(X) = p(1-p)$

2. Binomial distribution

Let the random variable X follow a binomial distribution with parameters n and p. That is, $X \sim B(n, p)$.

Each independent experiment X_i has a $(0-1)$ distribution.

$$P\{X_i = 1\} = p, \qquad P\{X_i = 0\} = 1 - p$$

$$E(X_i) = p, \qquad \mathrm{Var}(X_i) = p(1-p)$$

The probability distribution of X is

$$P\{X = k\} = \binom{n}{k} p^k (1-p)^{n-k} \qquad X = \sum_{i=1}^{n} X_i = k = \begin{cases} 0 \\ 1 \\ 2 \\ \vdots \\ n \end{cases}$$

So $$E(X^L) = \sum_{k=1}^{\infty} x_k^{\ L} p_k = \sum_{k=1}^{n} k^L \binom{n}{k} p^k (1-p)^{n-k} \qquad L = 1, 2, \ldots$$

The calculation is complicated if we use the definition of mathematical expectation. However, we can calculate it in another way according to the properties of mathematical expectation and variance.

Because $$X = \sum_{i=1}^{n} X_i$$

Take mathematical expectation on both sides.

$$E(X) = E(\sum_{i=1}^{n} X_i) = \sum_{i=1}^{n} E(X_i) = nE(X_i) = np$$

Take variance on both sides.

$$\text{Var}(X) = \text{Var}(\sum_{i=1}^{n} X_i) = \sum_{i=1}^{n} \text{Var}(X_i) = n\text{Var}(X_i) = np(1-p)$$

That is, $E(X) = np$, $\text{Var}(X) = np(1-p)$

3. Poisson distribution

Mathematical expectation

$$E(X) = \sum_{k=1}^{\infty} x_k p_k = \sum_{k=0}^{\infty} k \frac{\lambda^k e^{-\lambda}}{k!} = \lambda e^{-\lambda} \sum_{k=1}^{\infty} \frac{\lambda^{k-1}}{(k-1)!} = \lambda e^{-\lambda} \cdot e^{\lambda} = \lambda$$

We can get

$$E(X^2) = E[X(X-1) + X] = E[X(X-1)] + E(X)$$

$$= \sum_{k=0}^{\infty} k(k-1) \frac{\lambda^k e^{-\lambda}}{k!} + \lambda = \lambda^2 e^{-\lambda} \sum_{k=2}^{\infty} \frac{\lambda^{k-2}}{(k-2)!} + \lambda = \lambda^2 e^{-\lambda} \cdot e^{\lambda} + \lambda = \lambda^2 + \lambda$$

Therefore, the variance is

$$\text{Var}(X) = E(X^2) - [E(X)]^2 = \lambda^2 + \lambda - \lambda^2 = \lambda$$

That is, $E(X) = \lambda$, $\text{Var}(X) = \lambda$

4. Hypergeometric distribution

Let X have a hypergeometric distribution. The probability distribution of X is

$$P\{X = k\} = \frac{\binom{M}{k}\binom{N-M}{n-k}}{\binom{N}{n}} \qquad X = k = \begin{cases} 0 \\ 1 \\ 2 \\ \vdots \\ n \end{cases}$$

So $$E(X) = \sum_{k=1}^{\infty} x_k p_k = \sum_{k=0}^{n} k \cdot \frac{\binom{M}{k}\binom{N-M}{n-k}}{\binom{N}{n}} = \frac{nM}{N}$$

$$\mathrm{Var}(X) = E(X^2) - [E(X)]^2 = \sum_{k=0}^{n} k^2 \cdot \frac{\binom{M}{k}\binom{N-M}{n-k}}{\binom{N}{n}} - (\frac{nM}{N})^2$$

$$= \frac{nM}{N}(1-\frac{M}{N})(\frac{N-n}{N-1})$$

That is, $E(X) = \frac{nM}{N}$, $\mathrm{Var}(X) = \frac{nM}{N}(1-\frac{M}{N})(\frac{N-n}{N-1})$

Or $E(X) = np$, $\mathrm{Var}(X) = np(1-p)(\frac{N-n}{N-1})$

where $p = \frac{M}{N}$.

5. Exponential distribution

Let the random variable X have an exponential distribution with parameter λ, and the probability density of X is

$$f(x) = \begin{cases} \lambda e^{-\lambda x} & x > 0 \\ 0 & elsewhere \end{cases}$$

The mathematical expectation of X is

$$E(X) = \int_{-\infty}^{\infty} xf(x)dx = \int_{0}^{\infty} x\lambda e^{-\lambda x} dx = \frac{1}{\lambda}$$

The variance of X is

$$\text{Var}(X) = E(X^2) - [E(X)]^2 = \int_0^\infty x^2 \lambda e^{-\lambda x} dx - \frac{1}{\lambda^2} = \frac{1}{\lambda^2}$$

That is, $E(X) = \dfrac{1}{\lambda}$, $\text{Var}(X) = \dfrac{1}{\lambda^2}$

6. Uniform distribution

Let X be uniformly distributed on the interval [a, b], and the probability density of X is

$$f(x) = \begin{cases} \dfrac{1}{b-a} & a \le x \le b \\ 0 & elsewhere \end{cases}$$

The mathematical expectation of X is

$$E(X) = \int_{-\infty}^{\infty} x f(x) dx = \int_a^b x \frac{1}{b-a} dx = \frac{a+b}{2}$$

The variance of X is

$$\text{Var}(X) = E(X^2) - [E(X)]^2 = \int_a^b x^2 \frac{1}{b-a} dx - \left(\frac{a+b}{2}\right)^2 = \frac{(b-a)^2}{12}$$

That is, $E(X) = \dfrac{a+b}{2}$, $\text{Var}(X) = \dfrac{(b-a)^2}{12}$

7. Normal distribution

Let X be normally distributed with parameters μ and σ^2. That is, X ~ N (μ, σ^2). The probability density of X is

$$f(x) = \frac{1}{\sigma\sqrt{2\pi}} e^{-\frac{1}{2}\left(\frac{x-\mu}{\sigma}\right)^2} \qquad -\infty < x < \infty \qquad \sigma > 0$$

The mathematical expectation of X is

$$E(X) = \int_{-\infty}^{\infty} xf(x)dx = \int_{-\infty}^{\infty} x\frac{1}{\sigma\sqrt{2\pi}}e^{-\frac{1}{2}\left(\frac{x-\mu}{\sigma}\right)^2}dx = \mu$$

The variance of X is

$$\text{Var}(X) = \int_{-\infty}^{\infty} (x-\mu)^2 f(x)dx = \int_{-\infty}^{\infty} (x-\mu)^2\frac{1}{\sigma\sqrt{2\pi}}e^{-\frac{1}{2}\left(\frac{x-\mu}{\sigma}\right)^2}dx = \sigma^2$$

That is, $E(X) = \mu$, $\text{Var}(X) = \sigma^2$

In other words, the two parameters μ and σ^2 in the expression of the normal distribution's probability density are the mathematical expectation and variance of the random variable X. Therefore, the distribution of a normally distributed random variable X is completely determined by its mathematical expectation μ and variance σ^2. □

*Below, we introduce two important inequalities.

1. Let the random variable X ≥ 0. Then, for any positive number ε > 0, the following inequality is valid.

$$P\{X \geq \varepsilon\} \leq \frac{E(X)}{\varepsilon}$$

This inequality is called Markov's inequality.

Proof. For ε > 0, let

$$Z = \begin{cases} 1 & X \geq \varepsilon \\ 0 & otherwise \end{cases}$$

Since $0 \leq Z \leq 1$, $X \geq 0$, we have

$$Z \leq 1 \leq \frac{X}{\varepsilon} \quad \Rightarrow \quad Z \leq \frac{X}{\varepsilon}$$

Taking mathematical expectation on both sides gives

$$E(Z) \le \frac{1}{\varepsilon} E(X)$$

Because Z follows a (0 – 1) distribution so

$$E(Z) = 1 \cdot P\{Z = 1\} = P\{X \ge \varepsilon\}$$

Thus we have

$$P\{X \ge \varepsilon\} \le \frac{E(X)}{\varepsilon}$$

2. Let the random variable X have mathematical expectation $E(X) = \mu$ and variance $\mathrm{Var}(X) = \sigma^2$. Then, for any positive number $\varepsilon > 0$, the following inequality is valid.

$$P\{|X - \mu| \ge \varepsilon\} \le \frac{\sigma^2}{\varepsilon^2}$$

This inequality is called Chebyshev's inequality.

Proof. Method 1

Because $(X - \mu)^2 \ge 0$

Using Markov's inequality gives

$$P\{|X - \mu| \ge \varepsilon\} = P\{(X - \mu)^2 \ge \varepsilon^2\} \le \frac{E\left[(X - \mu)^2\right]}{\varepsilon^2} = \frac{\sigma^2}{\varepsilon^2}$$

Thus we have $P\{|X - \mu| \ge \varepsilon\} \le \frac{\sigma^2}{\varepsilon^2}$

Method 2. We only prove this for continuous random variables. Let the probability density of X be $f(x)$, then we have

$$P\{|X - \mu| \ge \varepsilon\} = \int_{|x-\mu|\ge\varepsilon} f(x)dx \le \int_{|x-\mu|\ge\varepsilon} \frac{|x - \mu|^2}{\varepsilon^2} f(x)dx \le \frac{1}{\varepsilon^2} \int_{-\infty}^{\infty} (x - \mu)^2 f(x)dx = \frac{\sigma^2}{\varepsilon^2}$$

where $\frac{(x-\mu)^2}{\varepsilon^2} \ge 1$, integral interval $\{|x - \mu| \ge \varepsilon\} \subseteq \{-\infty < x < \infty\}$.

Thus we have

$$P\{|X-\mu|\geq \varepsilon\}\leq \frac{\sigma^2}{\varepsilon^2}$$ □

Chebyshev's inequality can also be written in the following form.

$$P\{|X-\mu|<\varepsilon\}\geq 1-\frac{\sigma^2}{\varepsilon^2}$$

Or $$P\{|X-E(X)|<\varepsilon\}\geq 1-\frac{\mathrm{Var}(X)}{\varepsilon^2}$$

This inequality gives an estimation method of the probability of the event $\{|X-\mu|<\varepsilon\}$ when the distribution of the random variable X is unknown. For example, taking $\varepsilon = 3\sigma, 4\sigma$, we get

$$P\{|X-\mu|<3\sigma\}\geq 1-\frac{\sigma^2}{(3\sigma)^2}=\frac{8}{9}$$

$$P\{|X-\mu|<4\sigma\}\geq 1-\frac{\sigma^2}{(4\sigma)^2}=\frac{15}{16}$$

As a theoretical tool, Chebyshev's inequality has universal applications. If the absolute value is removed, the one-sided Chebyshev's inequality is

$$P\{X-\mu\geq \varepsilon\}\leq \frac{\sigma^2}{\sigma^2+\varepsilon^2}$$

$$P\{X-\mu\leq -\varepsilon\}\leq \frac{\sigma^2}{\sigma^2+\varepsilon^2}$$

Proof. Let $Y = X-\mu$

then $E(Y) = E(X) - \mu = 0$

$\mathrm{Var}(Y) = \mathrm{Var}(X) = \sigma^2 = E(Y^2)$

Let $b > 0$. And we realise

$$Y \geq \varepsilon \quad \Leftrightarrow \quad Y + b \geq \varepsilon + b$$

so $\quad P\{Y \geq \varepsilon\} = P\{Y + b \geq \varepsilon + b\} \leq P\{|Y + b| \geq \varepsilon + b\} = P\{(Y + b)^2 \geq (\varepsilon + b)^2\}$

Using Markov's inequality gives

$$P\{Y \geq \varepsilon\} \leq P\{(Y + b)^2 \geq (\varepsilon + b)^2\} \leq \frac{E\left[(Y+b)^2\right]}{(\varepsilon+b)^2} = \frac{\sigma^2 + b^2}{(\varepsilon+b)^2}$$

Since b is any positive number, when $b = \frac{\sigma^2}{\varepsilon}$, $\frac{\sigma^2 + b^2}{(\varepsilon+b)^2}$ obtains the minimum value.

Substituting $b = \frac{\sigma^2}{\varepsilon}$ into the inequality gives

$$P\{Y \geq \varepsilon\} \leq \frac{\sigma^2}{\sigma^2 + \varepsilon^2}$$

That is, $\quad P\{X - \mu \geq \varepsilon\} \leq \frac{\sigma^2}{\sigma^2 + \varepsilon^2}$

Similarly, we can get

$$P\{X - \mu \leq -\varepsilon\} \leq \frac{\sigma^2}{\sigma^2 + \varepsilon^2}$$

Example. Let the weekly output of a factory be a random variable X. Its mathematical expectation is E(X) = 150, and the variance is Var(X) = 420. What is the upper limit of the probability that the output this week will be at least 180?

Solution. Using the one-sided Chebyshev's inequality gives

$$P\{X \geq 180\} = P\{X - 150 \geq 30\} \leq \frac{420}{420 + 30^2} = \frac{7}{22} \approx 0.318$$

If we directly use Markov's inequality, we get

$$P\{X \geq 180\} \leq \frac{E(X)}{180} = \frac{5}{6} \approx 0.833$$

This upper limit is relatively large, much larger than the upper limit obtained using the one-sided Chebyshev's inequality. The smaller the upper limit, the stronger the conclusion. If the upper limit is 1, this conclusion makes no sense.

4.4. Examples

Example 1. Let X represent the number that appears when rolling a die. Find E(X) and Var(X).

Solution. Because the number that appears when rolling a die is

$$X = \begin{cases} 1 \\ 2 \\ \vdots \\ 6 \end{cases}$$

They are equally likely to appear.

$$P(1) = P(2) = \ldots = P(6) = \frac{1}{6}$$

Therefore we get

$$E(X) = \sum_{k=1}^{\infty} x_k p_k = 1\times\frac{1}{6} + 2\times\frac{1}{6} + 3\times\frac{1}{6} + 4\times\frac{1}{6} + 5\times\frac{1}{6} + 6\times\frac{1}{6} = \frac{7}{2}$$

$$E(X^2) = \sum_{k=1}^{\infty} x_k^2 p_k = 1^2\times\frac{1}{6} + 2^2\times\frac{1}{6} + 3^2\times\frac{1}{6} + 4^2\times\frac{1}{6} + 5^2\times\frac{1}{6} + 6^2\times\frac{1}{6} = \frac{91}{6}$$

Hence $\quad \text{Var}(X) = E(X^2) - [E(X)]^2 = \frac{91}{6} - \left(\frac{7}{2}\right)^2 = \frac{35}{12}$

That is, $\quad E(X) = \frac{7}{2}, \quad \text{Var}(X) = \frac{35}{12}$

Example 2. 100 trees form a circle. 25 golden squirrels are living on these trees. Find the mathematical expectation of the number of golden squirrels living on 10 randomly adjacent trees.

Solution. Let every golden squirrel be

$$X_i = \begin{cases} 1 & \textit{The ith golden squirrel on 10 adjacent trees} \\ 0 & \textit{otherwise} \end{cases} \quad i = 1, 2, \ldots, 25$$

The probability distribution of X_i is

$$P\{X_i = 1\} = \frac{10}{100} = \frac{1}{10}$$

Because X_i obeys a (0 – 1) distribution, so

$$E(X_i) = 1 \cdot P\{X_i = 1\} = \frac{1}{10}$$

Let the number of golden squirrels living on 10 randomly adjacent trees be

$$X = \sum_{i=1}^{25} X_i$$

So, the mathematical expectation is

$$E(X) = \sum_{i=1}^{25} E(X_i) = 25 \times \frac{1}{10} = 2.5$$

Example 3. 165 students from one school travelled in 4 buses to a sports competition. The first bus had 38 students. The second bus has 40 students. The third bus had 42 students. The fourth bus had 45 students. After arriving at the destination, one student is randomly selected from the 165 students. Let X represent the number of students on the bus with which the randomly selected student travelled. Find $E(X)$.

Solution. It is equally possible to select one student from 165 students randomly. The number of students on the bus taken by the selected student is

$$X = \begin{cases} 38 \\ 40 \\ 42 \\ 45 \end{cases}$$

The probabilities that the selected student belongs to a bus with 38, 40, 42 and 45 people are, respectively

$$P\{X=38\} = \frac{\binom{38}{1}}{\binom{165}{1}} = \frac{38}{165}, \qquad P\{X=40\} = \frac{\binom{40}{1}}{\binom{165}{1}} = \frac{40}{165}$$

$$P\{X=42\} = \frac{\binom{42}{1}}{\binom{165}{1}} = \frac{42}{165}, \qquad P\{X=45\} = \frac{\binom{45}{1}}{\binom{165}{1}} = \frac{45}{165}$$

So

$$E(X) = \sum_{k=1}^{\infty} x_k p_k = 38\times\frac{38}{165} + 40\times\frac{40}{165} + 42\times\frac{42}{165} + 45\times\frac{45}{165} = \frac{6833}{165} \approx 41 \text{ students}$$

That is, $E(X) \approx 41$ students

Example 4. An airport bus departs from the airport with 25 passengers. There are 12 bus stops where passengers can get off. If it arrives at a bus stop and no passenger gets off, the bus will not stop. Let X represent the number of stops that the bus will stop. Find $E(X)$. (Let it be equally possible for each passenger to get off at each bus stop. And whether each passenger gets off is independent of each other).

Solution. Let every bus stop be

$$X_i = \begin{cases} 1 & \textit{Someone gets off (bus stops)} \\ 0 & \textit{No one gets off (bus doesn't stop)} \end{cases}$$

At the ith bus stop, the probability of any passenger getting off is $\frac{1}{12}$ and the probability of not getting off is $\frac{11}{12}$.

Because

$$P\{\text{Someone gets off}\} + P\{\text{No one gets off}\} = 1$$

$$P\{\text{No one gets off}\} = P\{\text{None of the 25 passengers gets off}\} = \left(\frac{11}{12}\right)^{25}$$

so
$$P\{\text{Someone gets off}\} = 1 - P\{\text{No one gets off}\} = 1-\left(\frac{11}{12}\right)^{25}$$

Hence we have

$$P\{X_i = 1\} = 1-\left(\frac{11}{12}\right)^{25}, \qquad P\{X_i = 0\} = \left(\frac{11}{12}\right)^{25}$$

Since X_i follows a $(0-1)$ distribution so

$$E(X_i) = 1 \cdot P\{X_i = 1\} = 1-\left(\frac{11}{12}\right)^{25} \qquad i = 1, 2, \ldots, 12$$

Because the number of stops is

$$X = \sum_{i=1}^{12} X_i$$

Therefore, we have

$$E(X) = \sum_{i=1}^{12} E(X_i) = 12\left[1-\left(\frac{11}{12}\right)^{25}\right] \approx 10.6$$

Example 5. Eight hunters wait for a flock of wild ducks to fly by. When a flock of 15 wild ducks flies over the heads of the hunters, the 8 hunters randomly target a wild duck and shoot simultaneously. Let the probability that each hunter hits a wild duck be p. Find the mathematical expectation of the number of wild ducks that escape.

Solution. Let every wild duck be

$$X_i = \begin{cases} 1 & \textit{The } i^{th} \textit{ wild duck escapes.} \\ 0 & \textit{otherwise} \end{cases} \qquad i = 1, 2, \ldots, 15$$

Because whether each hunter hits the ith duck is independent of the other, the probability is

$$P\{\text{The } i\text{th duck is selected and hit by a hunter occur simultaneously}\} = \frac{1}{15}p$$

$$P\{\text{The } i\text{th wild duck escapes the shot of a hunter}\} = 1 - \frac{1}{15}p$$

So $$P\{X_i = 1\} = P\{\text{escapes the shooting of 8 hunters}\} = \left(1-\frac{1}{15}p\right)^8$$

Because X_i obeys a $(0-1)$ distribution, so

$$E(X_i) = 1 \cdot P\{X_i = 1\} = \left(1-\frac{1}{15}p\right)^8$$

Let the number of wild ducks escape be

$$X = \sum_{i=1}^{15} X_i$$

So, the mathematical expectation of the number of wild ducks escape is

$$E(X) = \sum_{i=1}^{15} E(X_i) = 15\left(1-\frac{1}{15}p\right)^8$$

Example 6. There are n keys that look the same. Only one of them can open the lock on the door. Use them to try to open the lock on the door. Let obtaining each key be equally possible. If the key is removed after being tried once, find the mathematical expectation of the number of times X the lock on the door has been tried.

Solution. This is sampling without replacement. Let each unlock trial be

$$
X_k = \begin{cases} 1 & unlock \qquad P\{X_k = 1\} = \dfrac{\binom{1}{1}\binom{n-1}{k-1}}{\binom{n}{k}} = \dfrac{k}{n} \\ 0 & doesn't \quad unlock \qquad P\{X_k = 0\} = 1 - \dfrac{k}{n} = \dfrac{n-k}{n} \end{cases} \qquad k = 1, 2, \ldots, n
$$

Because X_k follows a (0 – 1) distribution so we have

$$
E(X_k) = 1 \cdot P\{X_k = 1\} = \frac{k}{n}
$$

The number of unlock attempts is

$$
X = \sum_{k=1}^{n} X_k
$$

So $E(X) = \sum_{k=1}^{n} E(X_k) = \sum_{k=1}^{n} \frac{k}{n} = \frac{1}{n} \cdot \frac{(1+n)n}{2} = \frac{n+1}{2}$ (Sum of an arithmetic sequence)

That is, the mathematical expectation of the number of unlock attempts X is

$$
E(X) = \frac{n+1}{2}
$$

Example 7. Throw n balls (numbered 1 to n) randomly into n boxes (numbered 1 to n). If a ball is thrown into a box with the same number as the ball, it is called a pairing. Let X be the total number of pairings. Find E(X).

Solution. Let each time a ball is thrown into the box be

$$
X_i = \begin{cases} 1 & ball \quad thrown \quad \text{into} \quad a \quad same-numbered \quad box \quad P\{X_i = 1\} = \dfrac{\binom{1}{1}}{\binom{n}{1}} = \dfrac{1}{n} \\ 0 & ball \quad thrown \quad \text{into} \quad a \quad different-numbered \quad box \quad P\{X_i = 0\} = 1 - \dfrac{1}{n} \end{cases}
$$

$$
i = 1, 2, \ldots, n
$$

Because X_i follows a (0 – 1) distribution so we have

$$E(X_i) = 1 \cdot P\{X_i = 1\} = \frac{1}{n}$$

The total number of pairings is $\quad X = \sum_{i=1}^{n} X_i$

So $$E(X) = \sum_{i=1}^{n} E(X_i) = n \cdot \frac{1}{n} = 1$$

That is, the mathematical expectation of the total number of pairings X is

$$E(X) = 1$$

Example 8. Let the probability densities of random variables X_1 and X_2 be respectively

$$f_1(x) = \begin{cases} 2e^{-2x} & x > 0 \\ 0 & elsewhere \end{cases} \qquad f_2(x) = \begin{cases} 4e^{-4x} & x > 0 \\ 0 & elsewhere \end{cases}$$

(1) Find $E(X_1 + X_2)$, $E(2X_1 - 3X_2^2)$.

(2) Let X_1, X_2 be independent of each other. Find $E(X_1X_2)$.

Solution. First, find

$$E(X_1) = \int_0^{\infty} x \cdot 2e^{-2x} dx = \left[-xe^{-2x}\right]_0^{\infty} + \int_0^{\infty} e^{-2x} dx$$

$$= \int_0^{\infty} e^{-2x} dx = \left[-\frac{e^{-2x}}{2}\right]_0^{\infty} = \frac{1}{2}$$

Similarly, we get

$$E(X_2) = \int_0^{\infty} x \cdot 4e^{-4x} dx = \frac{1}{4}$$

$$E(X_2^2) = \int_0^{\infty} x^2 \cdot 4e^{-4x} dx = \left[-x^2 e^{-4x}\right]_0^{\infty} + \int_0^{\infty} 2xe^{-4x} dx = \int_0^{\infty} 2xe^{-4x} dx = \frac{1}{8}$$

Therefore $$E(X_1 + X_2) = E(X_1) + E(X_2) = \frac{1}{2} + \frac{1}{4} = \frac{3}{4}$$

$$E(2X_1 - 3X_2^2) = 2E(X_1) - 3E(X_2^2) = 2 \cdot \frac{1}{2} - 3 \cdot \frac{1}{8} = \frac{5}{8}$$

$$E(X_1X_2) = E(X_1)E(X_2) = \frac{1}{2} \cdot \frac{1}{4} = \frac{1}{8}$$

That is,

(1) $E(X_1 + X_2) = \frac{3}{4}$ $E(2X_1 - 3X_2^2) = \frac{5}{8}$

(2) $E(X_1X_2) = \frac{1}{8}$

Example 9. The lifetime X (in years) of a certain piece of equipment produced by a factory obeys an exponential distribution. The probability density is

$$f(x) = \begin{cases} \frac{1}{3}e^{-\frac{x}{3}} & x > 0 \\ 0 & elsewhere \end{cases}$$

The factory states that the equipment sold can be returned if damaged within one year of sale. If the factory sells one piece of equipment for €500, the profit is €200. Find the mathematical expectation (i.e., the average profit) of the factory's profit from selling a piece of equipment.

Solution. The probability that the equipment will be damaged within one year is

$$P\{0 < X < 1\} = \int_0^1 \frac{1}{3}e^{-\frac{x}{3}}dx = \left[-e^{-\frac{x}{3}}\right]_0^1 = 1 - e^{-\frac{1}{3}}$$

The probability that the equipment will not be damaged within one year is

$$p = 1 - P\{0 < X < 1\} = 1 - \left(1 - e^{-\frac{1}{3}}\right) = e^{-\frac{1}{3}}$$

Because the selling price is €500 and the profit is €200, so

Production cost = €500 – €200 = €300

The selling price of one piece of equipment

$$= \begin{cases} 500 & undamaged & p = e^{-\frac{1}{3}} \\ 0 & damaged & q = 1 - p = 1 - e^{-\frac{1}{3}} \end{cases}$$

average selling price = $0 \cdot (1-p) + 500p = 500e^{-\frac{1}{3}}$

So, the mathematical expectation of profit from selling a piece of equipment μ (i.e. average profit) is

$$\mu = 500e^{-\frac{1}{3}} - 300 \approx 58.27$$

That is, $\mu = €58.27$

Example 10. The defective rate of a particular product is 0.1. The inspector inspects 5 times a day. 10 products are randomly selected for inspection each time. If the inspector finds that the number of defective products exceeds 1, he will adjust the equipment. Let X represent the number of times the equipment is adjusted daily. Find *E*(X) (whether the product is defective or not is independent of each other).

Solution. Let each product be

$$\mathrm{Y}_i = \begin{cases} 1 & defective & p = 0.1 \\ 0 & non-defective & q = 1 - p = 0.9 \end{cases} \qquad i = 1, 2, \ldots, 10$$

The number of defective products among the 10 products inspected is

$$\mathrm{Y} = \sum_{i=1}^{10} \mathrm{Y}_i \sim \mathrm{B}\,(10, 0.1) \qquad \text{follows a binomial distribution.}$$

So, the probability that the number of defective products is more than 1 is

$$\mathrm{P}\{\mathrm{Y} > 1\} = 1 - [\mathrm{P}\{\mathrm{Y} = 0\} + \mathrm{P}\{\mathrm{Y} = 1\}]$$

$$= 1 - \left[\binom{10}{0} 0.1^0 0.9^{10} + \binom{10}{1} 0.1^1 0.9^{10-1}\right] \approx 0.2639$$

Let the equipment be adjusted after each inspection be

$$X_i = \begin{cases} 1 & adjust & p = 0.2639 \\ 0 & doesn't \quad adjust & q = 1 - p = 0.7361 \end{cases} \qquad i = 1, 2, 3, 4, 5$$

The number of times the equipment is adjusted in a day is

$$X = \sum_{i=1}^{5} X_i \sim B(5, 0.2639) \qquad \text{follows a binomial distribution.}$$

Hence
$$P\{X = 0\} = \binom{5}{0} 0.2639^0 0.7361^5 \approx 0.2161$$

$$P\{X = 1\} = \binom{5}{1} 0.2639^1 0.7361^{5-1} \approx 0.3874$$

$$P\{X = 2\} = \binom{5}{2} 0.2639^2 0.7361^{5-2} \approx 0.2778$$

$$P\{X = 3\} = \binom{5}{3} 0.2639^3 0.7361^{5-3} \approx 0.0996$$

$$P\{X = 4\} = \binom{5}{4} 0.2639^4 0.7361^{5-4} \approx 0.0179$$

$$P\{X = 5\} = \binom{5}{5} 0.2639^5 0.7361^{5-5} \approx 0.00128$$

Therefore

$$E(X) = \sum_{k=1}^{\infty} x_k p_k$$

$$= 0 \times 0.2161 + 1 \times 0.3874 + 2 \times 0.2778 + 3 \times 0.0996 + 4 \times 0.0179 + 5 \times 0.00128 \approx 1.32$$

That is, mathematical expectation (average number of adjustments) is

$$E(X) \approx 1.32$$

Example 11*. It is known that the average number of white blood cells X per millilitre in the blood of average adult men is 7300. The standard deviation is 700.

Use Chebyshev's inequality to estimate the probability p that the number of white blood cells per millilitre X is between 5500 and 9100.

Solution. Because the average is $E(X) = 7300$ and standard deviation $\sqrt{\text{Var}(X)} = 700$.

$$P\{5500 < X < 9100\} = P\{5500 - 7300 < X - 7300 < 9100 - 7300\}$$
$$= P\{-1800 < X - 7300 < 1800\}$$
$$= P\{|X - 7300| < 1800\}$$

From Chebyshev's inequality

$$P\{|X - E(X)| < \varepsilon\} \geq 1 - \frac{\text{Var}(X)}{\varepsilon^2}$$

we get $$P\{|X - 7300| < 1800\} \geq 1 - \frac{700^2}{1800^2} = \frac{275}{324}$$

That is, the probability that the number of white blood cells per millilitre X is between 5500 and 9100 is $P\{5500 < X < 9100\} \geq \frac{275}{324}$.

Exercise 4

1. A discrete random variable has the following probability distribution.

X	-1	0	1	2
p_k	0.2	0.1	0.4	0.3

Find **a.** $E(X)$ **b.** $E(X^2)$ **c.** $\text{Var}(X)$

d. $E(2X^2 + 4X - 3)$ **e.** $\text{Var}(-8X - 5)$

f. $Y = (X + 1)^2$

i. Find $E(Y)$; **ii.** Find $\text{Var}(Y)$.

2. A discrete random variable has the following probability distribution.

X	-2	0	3
p_k	0.4	0.5	0.1

Find **a.** E(X) **b.** Var(X) **c.** E(-X – 4) **d.** Var(-X – 4)

e. $Y = \dfrac{1}{X+3}$

i. Find E(Y); **ii.** Find Var(Y).

3. Let the random variable X ~ B (n, p). Given that E(X) = 10, Var(X) = 3.75, find the values of n and p.

4. The random variable X has the following probability density.

$$f(x) = \begin{cases} kx & 0 < x < 3 \\ 0 & elsewhere \end{cases}$$

Find

a. the value of k.

b. E(X)

c. Var(X)

d. the mean and variance of the random variable Y = -4X + 2.

5. The time to wait T minutes for service at a bank has the following probability density.

$$f(t) = \begin{cases} \frac{2}{15}t & 0 \le t \le 3 \\ -\frac{1}{5}t + 1 & 3 < t \le 5 \\ 0 & elsewhere \end{cases}$$

a. What is the median waiting time? That is, for what time t is the probability of waiting less than t minutes equal to $\frac{1}{2}$?

b. What is the mean waiting time?

c. What is the standard deviation of the waiting time?

6. Let the probability density of a random variable X be

$$f(x) = ax^2 + bx + c \qquad 0 < x < 1$$

Given that $E(X) = 0.5$, $Var(X) = 0.15$, find the values of constants a, b and c.

7. A and B play a game of tossing a fair die once. A will win 1 euro if a 6 appears. Otherwise, A wins nothing. Find the mean and variance of A's winning.

8. Let the probability of death of a certain bird at age 4 be 0.003. Find the mean and variance of the number of deaths among 25,000 birds at age 4.

9. A company's products are graded into 5 classes: class 1, class 2,…, and class 5 (in decreasing order of quality). The proportion of products in each class and their respective profits are shown in the following table.

X	class 1	class 2	class 3	class 4	class 5
p	$\frac{8}{20}$	$\frac{6}{20}$	$\frac{3}{20}$	$\frac{2}{20}$	$\frac{1}{20}$
profit (in euros)	1.3	1.1	0.8	0	-2

Find the mean and variance of profit of a product randomly selected from all company products.

10. 0.1% of the bolts produced by a machine are defective.

a. What is the maximum number of bolts that can be packed in a box if the probability of one or more defective bolts in the box is no more than 0.2?

b. What are the expected value and standard deviation of the number of defectives in a box of size found in **a.**?

c. Now, 250 bolts are packed in a box. 10 boxes are randomly selected for inspection. What is the expected number of boxes with no defective bolts?

11. In a game, 3 balls are drawn without replacement from an urn containing 3 red and 4 white balls. The player receives one euro if the drawing includes 2 or more red balls. Otherwise, the player loses fifty cents. What is the mathematical expectation of the player's gain?

12. The number of accidents that occur at a particular intersection on Monday mornings is 0, 1, 2, 3 with corresponding probabilities 0.95, 0.03, 0.01, 0.01. Find

a. the expected number of accidents on one Monday morning.

b. the expected number of accidents during 200 Monday mornings.

13. If Var(X) = 0.6, find

a. Var(10X) **b.** Var(3X + 2)

14. In an amusement park, players can pay 20 euros per hour to fish 20 shrimps with a fishing rod from a pond. In the pond, large shrimps account for 30%, medium shrimps account for 30%, extra large shrimps account for 5%, and small shrimps account for 35%. The average weights of large, medium, extra-large and small shrimps are 80 g, 50 g, 120 g and 30 g, respectively. What is the expected weight of a player's catch in one hour with 20 shrimps?

15. A box has 5 stopwatches, three of which are defective. A student needs a stopwatch for a physics experiment, so she takes stopwatches at random from the box, one after the other without replacement, until she gets a stopwatch that is not defective. What is the expected number of stopwatches the student takes from the box?

16. In a casino game roulette, a wheel with 35 slots is spun and a marble drops randomly into one of the slots. There are 16 red slots, 16 black slots and 3 green slots. If a player bets €50 on red, the player will win €50 if the marble drops in a red slot. Otherwise, the player loses €50. Calculate the expected winning of the player.

17. In a company's New Year party, one lucky game works like this: Box A has 3 silver and 5 black balls. Box B has 2 silver balls and 2 black balls. The game consists of two stages. First, throw a fair die. The player gets box A if the number 1, 2, or 3 appears. The player box B if the number 4, 5 or 6 appears. Then, the player randomly chooses two balls from the box without replacement. The player wins €200 for each silver ball drawn.

- **a.** Let X be the number of silver balls drawn. Find the probability distribution of X.
- **b.** What is the expected amount of money that a player wins?

18. An urn contains 5 balls, numbered 1 to 5. Take 3 balls at random from the urn without replacement. Let X be the largest number in the drawing. Find the mean and variance of X.

19*. X is a random variable with a mean of 6 and a standard deviation of 2. Use Chebyshev's inequality to estimate the probability that X is between 0 and 12.

20*. The maximum wind speed at a town on a randomly selected day is a random variable X with a mean of 8 m/s and a standard deviation of 2 m/s. If X exceeds 20 m/s, the town will cut off electricity. Use Chebyshev's inequality to find a bound for the probability of electricity cut-off for a randomly selected day.

Chapter 5* The law of large numbers and the central limit theorem

5.1. The law of large numbers

In Chapter 2, we mentioned that the relative frequency of the occurrence of events is stable. As the number of trials increases, the relative frequency of the occurrence of events gradually stabilises at a specific constant. In practice, people also realise that the arithmetic average of a large number of measured values also has stability. This stability is the objective background of the law of large numbers that needs to be discussed in this section.

Theorem 1 (a special case of Chebyshev's theorem)

Let the random variables $X_1, X_2, \ldots, X_n, \ldots$ be independent of each other and have the same mathematical expectation $E(X_i) = \mu$ and $Var(X_i) = \sigma^2$ $(i = 1, 2, \ldots)$. The arithmetic average of the first n random variables is

$$\bar{X} = \frac{1}{n}\sum_{i=1}^{n} X_i$$

Then, for any positive number $\varepsilon > 0$, we have

$$\lim_{n\to\infty} P\{|\bar{X} - \mu| < \varepsilon\} = 1$$

Therefore we get

$$\lim_{n\to\infty} \bar{X} = \mu$$

Proof. Because $E(\bar{X}) = \frac{1}{n}\sum_{i=1}^{n} E(X_i) = \frac{1}{n}\cdot n\mu = \mu$

$$Var(\bar{X}) = \frac{1}{n^2}\sum_{i=1}^{n} Var(X_i) = \frac{1}{n^2}\cdot n\sigma^2 = \frac{\sigma^2}{n}$$

Let $X = \bar{X}$. Substituting into Chebyshev's inequality, we get

$$P\{|\bar{X} - E(\bar{X})| < \varepsilon\} \geq 1 - \frac{Var(\bar{X})}{\varepsilon^2}$$

Let $n \to \infty$, then

$$P\{|\lim_{n\to\infty} \bar{X} - \mu| < \varepsilon\} \geq 1 - \frac{\lim_{n\to\infty} Var(\bar{X})}{\varepsilon^2}$$

where $\lim_{n\to\infty} Var(\bar{X}) = \lim_{n\to\infty} \frac{\sigma^2}{n} = 0$.

Note that the probability cannot be greater than 1, so

$$P\{|\lim_{n\to\infty} \bar{X} - \mu| < \varepsilon\} = 1$$

Because the given condition $\varepsilon > 0$ is a definite event so we have

$$P\{0 < \varepsilon\} = 1 = P\{|\lim_{n\to\infty} \bar{X} - \mu| < \varepsilon\}$$

Thus we have

$$\lim_{n\to\infty} \bar{X} = \mu \qquad \text{or} \qquad \lim_{n\to\infty} \bar{X} = E(X)$$

where $E(X) = \mu$, $\bar{X} = \bar{x}$, $X = x$. □

The result here is given by the probability method, which can be denoted as $\bar{X} \xrightarrow{P} \mu$. It means that $\bar{X}$ converges to μ according to probability.

Theorem 1 shows that when $n \to \infty$, the limit $\lim_{n\to\infty} \bar{X}$ becomes a constant μ. That is, $\lim_{n\to\infty} \bar{X} = \mu$. When n is relatively large, the arithmetic average is close to the mathematical expectation μ. That is, $\bar{X} \approx \mu$.

Theorem 2 (Bernoulli's theorem)

Let f_A be the number of times event A occurs in n independent repeated experiments. P(A) is the probability that event A occurs in each trial. Then, for any positive number ε > 0, we have

$$\lim_{n\to\infty} P\{\left|\frac{f_A}{n} - P(A)\right| < \varepsilon\} = 1$$

Thus we get

$$\lim_{n\to\infty} f_n(A) = \lim_{n\to\infty} \frac{f_A}{n} = P(A)$$

Proof. Let X_k denote the kth trial.

$$X_k = \begin{cases} 0 & A \quad doesn't \quad occur \\ 1 & A \quad occurs \end{cases} \qquad k = 1, 2, \ldots$$

Obviously, $\quad f_A = X_1 + X_2 + \ldots + X_n$

So $\quad \bar{X} = \frac{1}{n}\sum_{k=1}^{n} X_k = \frac{f_A}{n}$

Because each trial is independent, so $X_1, X_2, \ldots$ are independent of each other. And X_k follows a (0 – 1) distribution so we have

$$E(\bar{X}) = E(\frac{1}{n}\sum_{k=1}^{n} X_k) = E(X_k) = P(A)$$

$$Var(\bar{X}) = Var(\frac{1}{n}\sum_{k=1}^{n} X_k) = \frac{Var(X_k)}{n} = \frac{P(A)[1 - P(A)]}{n}$$

$$\lim_{n\to\infty} Var(\bar{X}) = 0$$

From Theorem 1, we have

$$\lim_{n\to\infty} P\{\left|\frac{f_A}{n} - P(A)\right| < \varepsilon\} = 1$$

Therefore $\quad \lim_{n\to\infty} f_n(A) = \lim_{n\to\infty} \frac{f_A}{n} = P(A)$

which can be denoted as $\quad f_n(A) \xrightarrow{P} P(A)$

Bernoulli's theorem shows that the relative frequency of the occurrence of the event $\lim_{n\to\infty} f_n(\mathrm{A}) = \lim_{n\to\infty} \frac{f_{\mathrm{A}}}{n}$ converges to P(A) according to probability. This theorem expresses the stability of relative frequency in a rigorous mathematical form. In practical applications, when the number of trials n is relatively large, the relative frequency of the occurrence of the event $f_n(\mathrm{A})$ can be used to replace the probability P(A) of the event.

Theorem 3 (Khinchin's theorem)

Let the random variables $X_1, X_2, \ldots, X_n, \ldots$ be independent of each other, follow the same distribution and have mathematical expectation $E(X_i) = \mu$ $(i = 1, 2, \ldots)$. Then, for any positive number $\varepsilon > 0$, we have

$$\lim_{n\to\infty} \mathrm{P}\{|\bar{\mathrm{X}} - \mu| < \varepsilon\} = 1$$

$$\lim_{n\to\infty} \bar{\mathrm{X}} = \mu$$

The proof here is similar to the proof of Theorem 1. The difference is that

$$\mathrm{Var}(\bar{\mathrm{X}}) = \frac{\sum_{i=1}^{n} \mathrm{Var}(\mathrm{X}_i)}{n^2} = \frac{\overline{\sigma^2}}{n} \neq \frac{\sigma^2}{n}$$

But $\lim_{n\to\infty} \mathrm{Var}(\bar{\mathrm{X}})$ is also equal to 0, so Theorem 3 is valid.

5.2. The central limit theorem

In objective reality, many random variables are formed by the combined influence of a large number of independent random variables. This kind of random variables approximately follows a normal distribution. This phenomenon is the objective background of the central limit theorem. This section introduces three commonly used central limit theorems.

Theorem 1 (Central limit theorem for independent and identical distributions)

Let the random variables $X_1, X_2, \ldots, X_n, \ldots$ be independent of each other, follow the same distribution and have mathematical expectation $E(X_i) = \mu$ and variance $\mathrm{Var}(X_i) = \sigma^2 \neq 0$ ($i = 1, 2, \ldots$), then for the random variable

$$Y_n = \frac{\sum_{i=1}^{n} X_i - E(\sum_{i=1}^{n} X_i)}{\sqrt{\mathrm{Var}(\sum_{i=1}^{n} X_i)}} = \frac{\sum_{i=1}^{n} X_i - n\mu}{\sqrt{n}\sigma}$$

its distribution function $F_n(x)$ for any x satisfies

$$\lim_{n\to\infty} F_n(x) = \lim_{n\to\infty} P\{\frac{\sum_{i=1}^{n} X_i - n\mu}{\sqrt{n}\sigma} \le x\} = \int_{-\infty}^{x} \frac{1}{\sqrt{2\pi}} e^{-\frac{y^2}{2}} dy \quad \text{[standard normal distribution]}$$

When n is relatively large

$$F_n(x) = P\{\frac{\sum_{i=1}^{n} X_i - n\mu}{\sqrt{n}\sigma} \le x\} \approx \int_{-\infty}^{x} \frac{1}{\sqrt{2\pi}} e^{-\frac{y^2}{2}} dy$$

Proof. Because

$$\lim_{n\to\infty} Y_n = \lim_{n\to\infty} \frac{\sum_{i=1}^{n} X_i - n\mu}{\sqrt{n}\sigma} = \lim_{n\to\infty} \frac{\bar{X} - \mu}{\sigma / \sqrt{n}} = \frac{0}{0} = y$$

That is, $\lim_{n\to\infty} Y_n = y$ (any number) $\quad -\infty < y < \infty$

The range of $\lim_{n\to\infty} Y_n$ conforms to the range of normal distribution so we have

$$P\{-\infty < \lim_{n\to\infty} Y_n < \infty\} = \int_{-\infty}^{\infty} f(y) dy = 1$$

Then, the probability density of $\lim_{n\to\infty} Y_n$ is

$$f(y) > 0, \quad -\infty < y < \infty$$

It conforms to the properties of normal probability density.

In addition, because

$$\lim_{n\to\infty} \bar{X} = E(X) = \mu$$

so when n is relatively large, we have

$$\bar{X} \approx \mu \ (\text{or} = \mu) \qquad \text{or} \qquad \bar{X} \approx \lim_{n\to\infty} \bar{X}$$

Because $\bar{X} \approx \mu$ (or $= \mu$)

$$\Rightarrow \text{relative error } \left|\frac{\bar{X} - \mu}{\mu}\right| \approx 0 \ (\text{or} = 0) \quad \Leftrightarrow \quad 0 \le \left|\frac{\bar{X} - \mu}{\mu}\right| \le 0.05$$

That is, $\mu - 0.05|\mu| \le \bar{X} \le \mu + 0.05|\mu|$

$\bar{X}$ takes values symmetrically around μ, which conforms to the properties of normal random variables. So we have

$$P\{\mu - 0.05|\mu| \le \bar{X} \le \mu + 0.05|\mu|\} = \int_{\mu-0.05|\mu|}^{\mu+0.05|\mu|} f(x)dx = 1$$

The probability density of $\bar{X}$ is

$$\begin{cases} f(x) > 0 & \mu - 0.05|\mu| \le x \le \mu + 0.05|\mu| \\ f(x) = 0 & \textit{elsewhere} \end{cases}$$

Because $\bar{X}$ conforms to the properties of normal random variables, in combination with the range of $\lim_{n\to\infty} Y_n$ is the normal range and $f(y)$ conforms to the properties of normal probability density, so

$$\bar{X} \approx \lim_{n\to\infty} \bar{X} \ \sim N\left(\mu, \lim_{n\to\infty}\frac{\sigma^2}{n}\right)$$

That is, $\bar{X}$ approximately follows the normal distribution $N\left(\mu, \frac{\sigma^2}{n}\right)$ $\left(\frac{\sigma^2}{n} \approx \lim_{n\to\infty}\frac{\sigma^2}{n}\right)$.

Therefore $$\frac{\bar{X} - \mu}{\sigma / \sqrt{n}} = Y_n \approx \lim_{n\to\infty} Y_n \sim N(0, 1)$$

According to the results derived above, we can get

$$P\{Y_n \le x\} \approx \lim_{n\to\infty} P\{Y_n \le x\} = \int_{-\infty}^{x} f(y)dy = \int_{-\infty}^{x} \frac{1}{\sqrt{2\pi}} e^{-\frac{y^2}{2}} dy$$

Hence we have

$$\lim_{n\to\infty} F_n(x) = \lim_{n\to\infty} P\{\frac{\sum_{i=1}^{n} X_i - n\mu}{\sqrt{n}\sigma} \le x\} = \int_{-\infty}^{x} \frac{1}{\sqrt{2\pi}} e^{-\frac{y^2}{2}} dy$$

When n is relatively large

$$F_n(x) = P\{\frac{\sum_{i=1}^{n} X_i - n\mu}{\sqrt{n}\sigma} \le x\} \approx \int_{-\infty}^{x} \frac{1}{\sqrt{2\pi}} e^{-\frac{y^2}{2}} dy$$

It can be known from here that $\sum_{i=1}^{n} X_i$ approximately follows the normal distribution

$N\left[E\left(\sum_{i=1}^{n} X_i\right), \mathrm{Var}\left(\sum_{i=1}^{n} X_i\right)\right] = N\,(n\mu, n\sigma^2)$.

Theorem 2 (Liapunov theorem)

Let the random variables $X_1, X_2, \ldots, X_n, \ldots$ be independent of each other. They have mathematical expectation $E(X_i) = \mu_i$ and variance $\mathrm{Var}(X_i) = \sigma_i^2 \ne 0$ ($i = 1, 2, \ldots$), denoted as $B_n^2 = \sum_{i=1}^{n} \sigma_i^2$. Then, for the random variable

$$Z_n = \frac{\sum_{i=1}^{n} X_i - E(\sum_{i=1}^{n} X_i)}{\sqrt{\mathrm{Var}(\sum_{i=1}^{n} X_i)}} = \frac{\sum_{i=1}^{n} X_i - \sum_{i=1}^{n} \mu_i}{B_n}$$

its distribution function $F_n(x)$ for any x satisfies

$$\lim_{n\to\infty} F_n(x) = \lim_{n\to\infty} P\{\frac{\sum_{i=1}^{n} X_i - \sum_{i=1}^{n} \mu_i}{B_n} \le x\} = \int_{-\infty}^{x} \frac{1}{\sqrt{2\pi}} e^{-\frac{z^2}{2}} dz$$

When n is relatively large

$$F_n(x) = P\{\frac{\sum_{i=1}^{n} X_i - \sum_{i=1}^{n} \mu_i}{B_n} \le x\} \approx \int_{-\infty}^{x} \frac{1}{\sqrt{2\pi}} e^{-\frac{z^2}{2}} dz$$

The proof is similar to that of Theorem 1 and is omitted here.

Theorem 1 and 2 show that under the conditions of the theorem, when n is relatively large, the random variables Y_n and Z_n approximately follow the normal distribution N (0, 1). This means that no matter what distribution each random variable X_i (i = 1, 2,...) follows, as long as n is relatively large, their sum will approximately follow a normal distribution.

In many problems, the random variable under consideration can be expressed as the sum of many independent random variables. For example, the electricity consumption of a city is the sum of the electricity consumption of a large number of users. The measurement error of a physical experiment is composed of many small errors. They approximately follow a normal distribution. This is a basic reason why normal random variables occupy an important position in probability theory.

In mathematical statistics, we will see that the central limit theorem is the theoretical basis for the statistical inference of large samples.

Another central limit theorem is introduced below. It is a special case of Theorem 1.

Theorem 3 (DeMoivre – Laplace theorem)

Let the random variable $X = \sum_{i=1}^{n} X_i$ follow a binomial distribution with parameters

(n, p). Then, for any x, we have

$$\lim_{n\to\infty} P\{\frac{X - np}{\sqrt{np(1-p)}} \le x\} = \int_{-\infty}^{x} \frac{1}{\sqrt{2\pi}} e^{-\frac{y^2}{2}} dy$$

When n is relatively large

$$P\{\frac{X - np}{\sqrt{np(1-p)}} \le x\} \approx \int_{-\infty}^{x} \frac{1}{\sqrt{2\pi}} e^{-\frac{y^2}{2}} dy$$

Proof Because $X = \sum_{i=1}^{n} X_i$ follows a binomial distribution, so we have

$$E(\sum_{i=1}^{n} X_i) = np, \qquad \text{Var}(\sum_{i=1}^{n} X_i) = np(1-p)$$

Therefore, from Theorem 1, we get

$$\lim_{n\to\infty} P\{\frac{X - np}{\sqrt{np(1-p)}} \le x\} = \lim_{n\to\infty} P\{\frac{\sum_{i=1}^{n} X_i - E(\sum_{i=1}^{n} X_i)}{\sqrt{\text{Var}(\sum_{i=1}^{n} X_i)}} \le x\} = \int_{-\infty}^{x} \frac{1}{\sqrt{2\pi}} e^{-\frac{y^2}{2}} dy$$

Example 1. A signal receiving adder receives 30 noise voltages V_i $(i = 1, 2, \ldots, 30)$ simultaneously. Let them be mutually independent random variables, and all follow a uniform distribution on (0, 12). Denote it as $V = \sum_{i=1}^{30} V_i$. Find an approximate value for $P\{V > 185\}$.

Solution. Because the probability density of V_i is

$$f(v) = \frac{1}{b-a} = \frac{1}{12-0} = \frac{1}{12}$$

So
$$E(V_i) = \int_{a}^{b} v f(v) dv = \int_{0}^{12} \frac{1}{12} v dv = 6$$

$$\mathrm{Var}(\mathrm{V}_i)=\int_a^b (v-\mu)^2 f(v)dv=\int_0^{12}\frac{1}{12}(v-6)^2 dv=12$$

From Theorem 1, the random variable

$$\mathrm{Z}=\frac{\sum_{i=1}^{n}\mathrm{V}_i-n\mu}{\sqrt{n}\sigma}=\frac{\sum_{i=1}^{30}\mathrm{V}_i-30\times 6}{\sqrt{30}\cdot\sqrt{12}}=\frac{\mathrm{V}-180}{6\sqrt{10}}$$

Because Z approximately ~ N (0, 1) and the probability density is $f(z)\approx\frac{1}{\sqrt{2\pi}}e^{-\frac{z^2}{2}}$

so
$$\mathrm{P}\{\mathrm{V}>185\}=\mathrm{P}\{\frac{\mathrm{V}-180}{6\sqrt{10}}>\frac{185-180}{6\sqrt{10}}\}=\mathrm{P}\{\frac{\mathrm{V}-180}{6\sqrt{10}}>\frac{\sqrt{10}}{12}\}$$

$$=\mathrm{P}\{\mathrm{Z}>\frac{\sqrt{10}}{12}\}\approx\int_{\frac{\sqrt{10}}{12}}^{\infty}\frac{1}{\sqrt{2\pi}}e^{-\frac{z^2}{2}}dz\ \approx 0.396$$

Hence, we have $\mathrm{P}\{\mathrm{V}>185\}\approx 0.396$.

Example 2. A ship is sailing in a certain sea area. It is known that every time a wave hits it, the probability of the pitch angle being greater than 3° is $p=\frac{1}{3}$. If the ship is hit by waves 80000 times, what is the probability that 26500 to 27000 of them have a pitch angle greater than 3°?

Solution. Represent each independent wave impact on the ship as X_i (i = 1, 2,…, 80000), then

$$\mathrm{X}_i=\begin{cases}0 & \{pitch\quad angle\ >3^\circ\}\quad doesn't\quad occur \qquad probability\ \ (1-p)=\frac{2}{3}\\ 1 & \{pitch\quad angle\ >3^\circ\}\quad occurs \qquad probability\ \ p=\frac{1}{3}\end{cases}$$

The number of times that the pitch angle is greater than 3° in 80000 wave impacts is represented by X, then $X=\sum_{i=1}^{80000} X_i$ follows a binomial distribution.

That is, $X \sim B\ (80000, \frac{1}{3})$. Its probability distribution is

$$P\{X=k\}=\binom{80000}{k}\left(\frac{1}{3}\right)^k\left(\frac{2}{3}\right)^{80000-k} \qquad k=0, 1,\ldots, 80000$$

The required probability is

$$P\{26500 \le X \le 27000\} = \sum_{k=26500}^{27000} P\{X=k\} = \sum_{k=26500}^{27000}\binom{80000}{k}\left(\frac{1}{3}\right)^k\left(\frac{2}{3}\right)^{80000-k}$$

Obviously, it is troublesome to calculate directly. We use Theorem 1 to find its approximation.

Because $E(X_i)=p$ $\qquad \mathrm{Var}(X_i)=p(1-p)$

$$X=\sum_{i=1}^{n} X_i \overset{approximately}{\sim} N\,[E(X), \mathrm{Var}(X)] = N\,[np, np(1-p)]$$

Therefore we have

$$P\{26500 \le X \le 27000\} = P\{\frac{26500-np}{\sqrt{np(1-p)}} \le \frac{X-np}{\sqrt{np(1-p)}} \le \frac{27000-np}{\sqrt{np(1-p)}}\}$$

$$\approx \int_{\frac{26500-np}{\sqrt{np(1-p)}}}^{\frac{27000-np}{\sqrt{np(1-p)}}} \frac{1}{\sqrt{2\pi}} e^{-\frac{z^2}{2}} dz = \Phi(\frac{27000-np}{\sqrt{np(1-p)}}) - \Phi(\frac{26500-np}{\sqrt{np(1-p)}})$$

Substituting $n = 80000$, $p=\frac{1}{3}$ gives

$$P\{26500 \le X \le 27000\} = \Phi(2.5) - \Phi(-1.25) \approx 0.888$$

Example 3. There is a shipment of wooden pillars, 85% of which is no less than 3 meters in length. Now, randomly select 100 wooden pillars from this batch. What is the probability that at least 20 are shorter than 3 meters?

Solution. Let each wooden pillar be

$$X_i = \begin{cases} 1 & less \quad than \quad 3 \quad meters & p = 1 - 0.85 = 0.15 \\ 0 & no \quad less \quad than \quad 3 \quad meters & q = 0.85 \end{cases}$$

$$i = 1, 2, \ldots, 100$$

Because X_i follows a $(0 - 1)$ distribution, so

$$E(X_i) = p = 0.15 \qquad \mathrm{Var}(X_i) = p(1 - p) = 0.15 \cdot 0.85 = 0.1275$$

Since $n = 100$ is relatively large, among 100 wooden pillars, the number of pillars that are shorter than 3 meters is

$$X = \sum_{i=1}^{100} X_i \overset{approximately}{\sim} N\left[E\left(\sum_{i=1}^{100} X_i\right), \mathrm{Var}\left(\sum_{i=1}^{100} X_i\right)\right] = N\,(100 \cdot 0.15,\ 100 \cdot 0.1275)$$

The probability that at least 20 of them are shorter than 3 meters is

$$P\{\sum_{i=1}^{100} X_i \geq 20\} \approx 0.0807 \qquad \text{(Calculate with a calculator.)}$$

Example 4. A company has a telephone switchboard. There are 150 telephone extensions. Let each telephone extension be independent of each other as to whether or not it is used. Let each extension have a 10% probability of using an external line at any time. How many external lines are needed at the switchboard to ensure, with a probability of no less than 95%, that each extension has an external line available when it wants to use an external line?

Solution. Let each extension use an external line to be

$$X_i = \begin{cases} 1 & use \quad external \quad line & p = 0.1 \\ 0 & not \quad \text{using} \quad external \quad line & q = 0.9 \end{cases} \qquad i = 1, 2, \ldots, 150$$

Because X_i follows a $(0 - 1)$ distribution, so

$$E(X_i) = p = 0.1 \qquad \mathrm{Var}(X_i) = p(1 - p) = 0.1 \cdot 0.9 = 0.09$$

Since $n = 150$ is relatively large, so the number of external lines required for 150 phones is

$$X = \sum_{i=1}^{150} X_i \overset{approximately}{\sim} N\left[E\left(\sum_{i=1}^{150} X_i\right), \text{Var}\left(\sum_{i=1}^{150} X_i\right)\right] = N\,(150\cdot 0.1,\ 150\cdot 0.09)$$

The switchboard needs to provide k external lines to guarantee their use with a probability of no less than 95%. That is,

$$0.95 \le P\{\sum_{i=1}^{150} X_i \le k\} = \Phi\left(\frac{k - 150\cdot 0.1}{\sqrt{150\cdot 0.09}}\right)$$

Hence $$\Phi\left(\frac{k - 150\cdot 0.1}{\sqrt{150\cdot 0.09}}\right)_{\min} = 0.95$$

Solving with a calculator gives

$$\frac{k - 150\cdot 0.1}{\sqrt{150\cdot 0.09}} \approx 1.645$$

So $$k_{\min} \approx 1.645\sqrt{150\cdot 0.09} + 150\cdot 0.1 \approx 21.04$$

That is, the switchboard must provide a minimum number of external lines $k_{\min} = 22$ or $k \ge 22$.

Example 5. A complex system consists of n components that function independently of each other. The reliability of each component (that is, the probability that the component will work) is 0.9. And at least 80% of the components must work for the entire system to work. What is the minimum value of n such that the system's reliability is no less than 0.95?

Solution. Let each component be

$$X_i = \begin{cases} 1 & \textit{component works} & p = 0.9 \\ 0 & \textit{component doesn't work} & q = 0.1 \end{cases} \qquad i = 1, 2, \ldots, n$$

Because X_i follows a $(0-1)$ distribution, so

$$E(X_i) = p = 0.9 \qquad \text{Var}(X_i) = p(1-p) = 0.9 \cdot 0.1 = 0.09$$

If n is relatively large, then the number of working components is

$$X = \sum_{i=1}^{n} X_i \overset{approximately}{\sim} N\left[E\left(\sum_{i=1}^{n} X_i\right),\ \text{Var}\left(\sum_{i=1}^{n} X_i\right)\right] = N\,(0.9n,\ 0.09n)$$

The reliability of the system is

$$0.95 \le P\{\sum_{i=1}^{n} X_i \ge 0.8n\} = \Phi(\infty) - \Phi\left(\frac{0.8n - 0.9n}{\sqrt{0.09n}}\right)$$

Rearranging gives

$$\Phi\left(\frac{0.8n - 0.9n}{\sqrt{0.09n}}\right) \le \Phi(\infty) - 0.95 = 1 - 0.95 = 0.05$$

That is,
$$\Phi\left(\frac{0.8n - 0.9n}{\sqrt{0.09n}}\right)_{\max} = \Phi\left(\frac{-\sqrt{n}}{3}\right)_{\max} = 0.05$$

Calculating with a calculator gives

$$\frac{-\sqrt{n}}{3} \approx -1.65$$

That is, $\qquad n_{\min} = 25 \qquad$ or $\qquad n \ge 25$

Therefore, when the number of components $n \ge 25$, the system's reliability is no less than 0.95.

Example 6. A pharmaceutical company claims that a drug produced by the company had a cure rate of 0.85 for treating a complex disease. A hospital inspector randomly examined 100 patients who had taken this drug. If more than 80 of them are cured, the claim is accepted, otherwise it is rejected.

(1) If the actual cure rate of this drug for this disease is 0.85, what is the probability of accepting the claim?

(2) If the actual cure rate of this drug for this disease is 0.75, what is the probability of accepting the claim?

Solution. (1) The actual cure rate is 0.85. Let everyone take the medicine be

$$X_i = \begin{cases} 1 & it \quad cures & p = 0.85 \\ 0 & it \quad doesn't \quad cure & q = 0.15 \end{cases} \qquad i = 1, 2, \ldots, 100$$

Because X_i follows a (0 – 1) distribution, so

$$E(X_i) = p = 0.85 \qquad \text{Var}(X_i) = p(1-p) = 0.85 \cdot 0.15 = 0.1275$$

Since $n = 100$ is relatively large, so among 100 people taking this medicine, the total number of people who can be cured is

$$X = \sum_{i=1}^{100} X_i \overset{approximately}{\sim} N\left[E\left(\sum_{i=1}^{100} X_i\right), \text{Var}\left(\sum_{i=1}^{100} X_i\right)\right] = N\,(100 \cdot 0.85,\ 100 \cdot 0.1275)$$

The probability that the total number of people cured X > 80, that is, the probability of accepting a cure rate of 0.85, is

$$P\{\sum_{i=1}^{100} X_i > 80\} \approx 0.919 \qquad \text{(Solve with a calculator)}$$

(2) The actual cure rate is 0.75. At this time

$$X = \sum_{i=1}^{100} X_i \overset{approximately}{\sim} N\,(100 \cdot 0.75,\ 100 \cdot 0.1875)$$

The probability that the total number of people cured X > 80, that is, the probability of accepting a cure rate of 0.85, is

$$P\{\sum_{i=1}^{100} X_i > 80\} \approx 0.124 \qquad \text{(Solve with a calculator)}$$

Example 7. Let X represent the number of times the head H comes up after tossing a coin 60 times. Find the probability that X = 30. Apply the normal approximation method and compare it with the exact solution.

Solution. Because

$$X = \sum_{i=1}^{60} X_i \sim B\left(60, \frac{1}{2}\right) \qquad \text{follows a binomial distribution.}$$

then $$E(X) = np = 60 \cdot \frac{1}{2} = 30, \quad Var(X) = np(1-p) = 60 \cdot \frac{1}{2} \cdot \frac{1}{2} = 15$$

Since $n = 60$ is relatively large so

$$X = \sum_{i=1}^{60} X_i \overset{approximately}{\sim} N\,[E(X), Var(X)] = N\,(30, 15)$$

However, the normal distribution is a continuous random variable and its probability $P\{X = 30\} = 0$. Therefore, the following continuity correction is required, namely

$$\{X = 30\} = \{30 \le X \le 30\} \approx \{30 - \frac{1}{2} \le X \le 30 + \frac{1}{2}\}$$

This way we get

$$P\{X = 30\} = P\{29.5 \le X \le 30.5\} \approx 0.1027 \quad [X \overset{approximately}{\sim} N\,(30, 15)]$$

The exact solution is

$$P\{X = 30\} = \binom{60}{30}\left(\frac{1}{2}\right)^{30}\left(1-\frac{1}{2}\right)^{60-30} \approx 0.1026$$

Example 8. 54% of residents in a city support banning smoking in public places. Randomly select n citizens.

(1) What is the probability that more than 50% of the people support this ban, if (a) $n = 25$ (b) $n = 100$ (c) $n = 225$?

(2) For this probability value to exceed 0.95, how big does n need to be?

Solution. (1) Let each resident be

$$X_i = \begin{cases} 1 & \textit{Support the ban} & p = 0.54 \\ 0 & \textit{Do not } \text{support } \textit{the ban} & 1-p = 0.46 \end{cases}$$

Because X_i has a $(0-1)$ distribution, so

$$E(X_i) = p = 0.54, \quad \mathrm{Var}(X_i) = p(1-p) = 0.54 \times 0.46$$

When n is relatively large, the number of people who support the ban is

$$X = \sum_{i=1}^{n} X_i \overset{approximately}{\sim} N\left[E\left(\sum_{i=1}^{n} X_i\right), \mathrm{Var}\left(\sum_{i=1}^{n} X_i\right)\right] = N\,(0.54n,\ 0.54 \times 0.46n)$$

so the probability that more than 50% of people support the ban is

$$P\{X > 0.5n\} = P\{\frac{X - 0.54n}{\sqrt{0.54 \times 0.46n}} > \frac{0.5n - 0.54n}{\sqrt{0.54 \times 0.46n}}\} = P\{\frac{X - 0.54n}{\sqrt{0.54 \times 0.46n}} > -0.08\sqrt{n}\}$$

$$\approx \Phi(\infty) - \Phi(-0.08\sqrt{n}) = \Phi(0.08\sqrt{n})$$

That is, $\quad P\{X > 0.5n\} \approx \Phi(0.08\sqrt{n})$

Therefore $\quad P\{X > 0.5n\} \approx \begin{cases} \Phi(0.4) \approx 0.655 & when \quad n = 25 \\ \Phi(0.8) \approx 0.788 & when \quad n = 100 \\ \Phi(1.2) \approx 0.885 & when \quad n = 225 \end{cases}$

(2) If this probability value exceeds 0.95, that is,

$$0.95 < P\{X > 0.5n\} \approx \Phi(0.08\sqrt{n})$$

then $\quad \Phi(0.08\sqrt{n}) > 0.95$

Solving with a calculator gives

$$0.08\sqrt{n} > 1.645$$

thus $\quad n > \left(\dfrac{1.645}{0.08}\right)^2 \approx 422.816$

That is, when $n \geq 423$ people, this probability value exceeds 0.95.

Exercise 5

1. The lifetime of a specific type of electronic component follows an exponential distribution with an average lifetime of 100 hours. Take randomly 20 of these types of components, and their lifetime are independent of each other. Find the probability that the total lifetime of these 20 components will exceed 2500 hours.

2. A particular calculator rounds each addition when performing addition. Let the errors in each rounding be independent of each other and follow a uniform distribution on (-0.5, 0.5).

a. If 200 addition operations are now to be performed, find the probability that the absolute value of the sum of the errors exceeds 10.

b. What is the maximum number of addition operations that can be performed to ensure that the probability that the absolute value of the sum of the errors does not exceed 12 is no less than 0.95?

3. In each experiment, the occurrence probability of event A is 0.3. Conduct the same experiment independently 1000 times. What is the probability of the occurrence times of event A being between 260 and 320 times?

4. In a college, there are 2000 students. It is set in the course selection system that at least 70 people must select a course before it can be started. Suppose the probability of each student choosing a particular course is 3%. Each student selects a course independently of each other. Find the probability that the course will be offered.

5. The weights of mechanical parts are random variables. They are independent of each other and follow the same distribution. They have a mathematical expectation of 0.4 kilograms and a standard deviation of 0.1 kilograms. What is the probability that the total weight of the 8000 parts will exceed 3220 kilograms?

6. A container contains 10,000 particles. The probability that a particle escapes from the container equals 0.0004. Assume that all the escapes are independent of each other. What is the probability that more than 5 escapes will occur?

Chapter 6 Population, random samples and statistics

6.1. Population and random samples

In mathematical statistics, we need to be concerned with a certain quantitative indicator of the research object. We call the total values of a certain quantitative indicator of the research object a population, and each element in the population is called an individual. Each individual x_i is a real number. We can use a set to express.

$$\{\text{Population}\} = \{x_1, x_2, \ldots, x_N\} = \{\text{X}\}$$

Therefore, we have $$\text{Population} = \text{X} = \begin{cases} x_1 \\ x_2 \\ \vdots \\ x_N \end{cases}$$

The population and the corresponding random variable X are equivalent. To distinguish the random variables in other cases, let (Population X) = X.

That is, Population = Population X

The population represents all individual data. For example, the lifespan of all bulbs produced by a factory is a population, and the lifespan of each bulb is an individual. The height of all male students in a school is a population, and the height of each male student is an individual.

The population is divided into finite population ($N < \infty$) and infinite population ($N = \infty$) according to the total number of individuals N the population contains. For example, in the population consisting of the lifespans of LED light bulbs produced in a factory in October, the total number of individuals N is the lifespans of LED light bulbs produced in October. And it is a finite population ($N < \infty$). The population of lifespans of all LED light bulbs produced by this factory is an infinite population ($N = \infty$), including the lifespans of LED light bulbs produced in the past and in the future. When the total number of individuals N contained in a finite population is huge, it can be approximately regarded as an infinite population, that is, $N \approx \lim_{N \to \infty} N$.

For example, let us examine the population of LED light bulb lifespans produced by a factory in October. We know the percentage of bulb lifespans X falling within each time interval. For example, bulb lifespans falling within 10000 – 15000 hours account for 85% of the total number of bulbs N, i.e., $P\{10000 \le X < 15000\} = 0.85$, and bulb lifespans falling within 15000 – 20000 hours account for 5% of the total number of bulbs N, i.e., $P\{15000 \le X < 20000\} = 0.05$ etc. Therefore, the values of bulb lifespans X have a specific distribution.

Generally speaking, the population we study, that is, a certain quantitative indicator X of the research object, has a certain objective distribution of its values. X is a random variable. Our study of the population is the study of the distribution of the corresponding random variable X. According to this, the distribution function and numerical characteristics of the population X represent the distribution function and numerical characteristics of the population, respectively. Henceforth, no distinction will be made between the population and the corresponding random variable X (i.e. population X).

To understand the nature of a population very clearly, the ideal way is to observe each individual one by one, but this is unrealistic. For example, if we want to study the life of light bulbs, because the life test is destructive, once we get all the test results, all the light bulbs will be burned out. We can only randomly select some bulbs from the entire batch of bulbs to do life tests, record their results, and then use these data to infer the life of the whole batch of bulbs.

For another example, for products such as rubber gaskets for beer bottle caps, it can be determined whether they are qualified through simple measurements, and the test is not destructive. However, due to the massive number of gaskets produced, measuring one by one takes a lot of workforce and time. Therefore, we still have to take a small number of gaskets for measurement and estimate the pass rate of the entire batch of gaskets based on the data obtained.

In general, we take a portion of individuals from the population for observation and then estimate the properties of the population based on the obtained data. The portion of individuals taken out is called a sample of the population.

Randomly selecting an individual from the population means making an observation (i.e. conducting an experiment) on population X and recording its results. We make n repeated, independent observations of the population X under the same conditions and record the results of the n observations, in the order of the trials, as $X_1, X_2, \ldots, X_n$. Since they are independent observations, it is reasonable to assume that $X_1, X_2, \ldots, X_n$ are random variables that are independent of each other and have the same distribution. The $X_1, X_2, \ldots, X_n$ obtained in this way are called a simple random sample from the population X, or a sample for short. n is called the sample size. Unless otherwise stated, the samples generally referred to are simple random samples.

When n observations have been completed, we get a set of real numbers $x_1, x_2, \ldots, x_n$. They are the observed values of the random variables $X_1, X_2, \ldots$, and X_n, respectively and are called sample values. Represent by a set.

$$\{\text{Sample}\} = \{X_1, X_2, \ldots, X_n\} = \{x_1, x_2, \ldots, x_n\}$$

For a finite population, simple random samples can be obtained by using sampling with replacement. However, sampling with replacement is inconvenient to use. When the total number of individuals N is much larger than the sample size n to be obtained, generally when $\frac{N}{n} \geq 10$, sampling without replacement can be treated as sampling with replacement in practice.

6.2. Numerical characteristics of population

It can be seen from the above that population = population X, so the numerical characteristics of the population = the numerical characteristics of the population X. At the same time, because $X = x$ (observation value). Generally, there is no distinction between random variable X and observation value x. They are equivalent.

I. Numerical Characteristics of Discrete Populations

(1) Mean of a discrete population, that is, the mathematical expectation $E(X)$

According to the average definition, we have

$$E(X) = \frac{\sum_{i=1}^{N} x_i}{\sum_{i=1}^{N} 1} = \frac{1}{N}\sum_{i=1}^{N} x_i$$

or

$$E(X) = \frac{\sum_{i=1}^{k} x_i f_i}{\sum_{i=1}^{k} f_i} = \frac{1}{N}\sum_{i=1}^{k} x_i f_i$$

or $$E(\mathrm{X}) = \frac{\sum_{i=1}^{\infty} x_i p_i}{\sum_{i=1}^{\infty} p_i} = \sum_{i=1}^{\infty} x_i p_i$$

where size $\mathrm{N} \le \infty$, $x_i f_i$ means that there are f_i pieces of x_i, $\sum_{i=1}^{k} f_i = \mathrm{N}$,

$\sum_{i=1}^{\infty} p_i = \sum_{i=1}^{\infty} \mathrm{P}\{\mathrm{X} = x_i\} = 1$.

(2) Variance of a discrete population, that is, variance Var(X)

Because $$\mathrm{Var(X)} = E\left[\left(\mathrm{X} - E(\mathrm{X})\right)^2\right] = E(\mathrm{X}^2) - \left[E(\mathrm{X})\right]^2$$

so $$\mathrm{Var(X)} = \frac{1}{\mathrm{N}} \sum_{i=1}^{\mathrm{N}} (x_i - \mu)^2$$

or $$\mathrm{Var(X)} = \frac{\sum_{i=1}^{k} (x_i - \mu)^2 f_i}{\sum_{i=1}^{k} f_i} = \frac{1}{\mathrm{N}} \sum_{i=1}^{k} (x_i - \mu)^2 f_i$$

or $$\mathrm{Var(X)} = \frac{1}{\mathrm{N}} \sum_{i=1}^{N} x_i^2 - \mu^2$$

or $$\mathrm{Var(X)} = \frac{\sum_{i=1}^{\infty} (x_i - \mu)^2 p_i}{\sum_{i=1}^{\infty} p_i} = \sum_{i=1}^{\infty} (x_i - \mu)^2 p_i$$

When finding the numerical characteristics of the population, all individual data are taken with sampling without replacement for calculation processing.

II. Numerical Characteristics of Continuous Populations

(1) Mean of a continuous population, that is, the mathematical expectation E(X)

According to the definition of average, we have

$$E(X)=\frac{\int_{-\infty}^{\infty} xf(x)dx}{\int_{-\infty}^{\infty} f(x)dx}=\int_{-\infty}^{\infty} xf(x)dx$$

where $f(x)$ is a probability density. $\int_{-\infty}^{\infty} f(x)dx=1$.

(2) Variance of a continuous population, that is, variance Var(X)

Because $\quad \text{Var}(X)= E\left[\left(X-E(X)\right)^2\right]=E(X^2)-\left[E(X)\right]^2$

so $\quad \text{Var}(X)=\frac{\int_{-\infty}^{\infty}(x-\mu)^2 f(x)dx}{\int_{-\infty}^{\infty} f(x)dx}=\int_{-\infty}^{\infty}(x-\mu)^2 f(x)dx$

or $\quad \text{Var}(X)=\int_{-\infty}^{\infty} x^2 f(x)dx-\mu^2$

6.3. Sample statistics

Samples are the basis for statistical inferences. In application, the samples themselves are not used directly. Instead, appropriate functions of the samples are constructed for different problems, and the functions of these samples are used for statistical inference.

Definition

Let $X_1, X_2, \ldots, X_n$ be a sample from the population and $g(X_1, X_2, \ldots, X_n)$ be a function of $X_1, X_2, \ldots, X_n$. If g is a continuous function and g does not contain any unknown parameters, then $g(X_1, X_2, \ldots, X_n)$ is said to be a statistic.

Let $x_1, x_2, \ldots, x_n$ be the sample value corresponding to the samples $X_1, X_2, \ldots, X_n$, then we say $g(x_1, x_2, \ldots, x_n)$ is the observed value of $g(X_1, X_2, \ldots, X_n)$. That is,

$$g(X_1, X_2, \ldots, X_n) = g(x_1, x_2, \ldots, x_n) \text{ observed value}$$

Several commonly used statistics are listed below. Let $X_1, X_2, \ldots, X_n$ be a sample from the population, and $x_1, x_2, \ldots, x_n$ be the observed value of this sample. No matter the population X is discrete or continuous, the sample statistic formula is the same.

Definition

Sample mean $\quad \bar{x} = \frac{1}{n}\sum_{i=1}^{n} x_i$

or $\quad \bar{x} = \frac{1}{n}\sum_{i=1}^{k} x_i f_i \quad (\sum_{i=1}^{k} f_i = n)$

Sample variance $\quad s_n^2 = \frac{1}{n}\sum_{i=1}^{n}(x_i - \bar{x})^2 = \frac{1}{n}\sum_{i=1}^{n} x_i^2 - \bar{x}^2$

Unbiased sample variance $\quad s_{n-1}^2 = \frac{1}{n-1}\sum_{i=1}^{n}(x_i - \bar{x})^2 = \frac{1}{n-1}\left(\sum_{i=1}^{n} x_i^2 - n\bar{x}^2\right)$

Sample standard deviation $\quad s_n = \sqrt{s_n^2} \quad$ or $\quad s_{n-1} = \sqrt{s_{n-1}^2}$

Sample statistics $\bar{x}$, s_n^2, s_{n-1}^2, $r, \ldots$ are all random variables, and they are all estimates.

Above, we only give statistics expressed in terms of observed values x_i. If necessary, we can substitute $x_i = X_i$ into the above formula to get the statistics represented by the random variable X_i $(i = 1, 2,\dots, n)$.

6.4. Definition of estimate

The distribution form of a population X is known, but one or many of its parameters (e.g. μ, σ^2, ρ,…) are unknown or impossible to get. The problem of estimating the value of the unknown parameter θ of the population using a sample of the population X is called a parameter estimation problem. For example,

$$\text{estimate} \quad \hat{\mu} = \bar{x}, \qquad \widehat{\sigma^2} = s_n^2, \qquad \hat{\rho} = r$$

$\uparrow$ estimator $\quad$ $\uparrow$ value of estimate

That is, population parameter estimate $\hat{\theta}$ = sample statistic

The samples are taken randomly. Hence, the results of sample statistics are random and can only be used as values of estimate.

Definition of an unbiased estimate

population parameter	$\theta = \mu, \sigma^2, \rho,\dots$
estimate of population parameter	$\hat{\theta} = \hat{\mu}, \widehat{\sigma^2}, \hat{\rho},\dots$
population parameter estimate	$\hat{\theta}$ = sample statistic

Since the error is

$$\varepsilon = \hat{\theta} - \theta$$

The mathematical expectation of error is

$$E(\varepsilon) = E(\hat{\theta}) - E(\theta) = E(\hat{\theta}) - \theta$$

where $E(\varepsilon)$ is called the systematic error.

When the system error is

$$E(\varepsilon) = E(\hat{\theta}) - \theta = 0$$

rearranging gives

$$E(\hat{\theta}) = \theta$$

An estimate $\hat{\theta}$ that satisfies this condition is called an unbiased estimate.

Example 1. Sample mean

$$\bar{x} = \frac{1}{n}\sum_{i=1}^{n} x_i$$

Because $$E(\bar{\mathrm{X}}) = \frac{1}{n}\sum_{i=1}^{n} E(\mathrm{X_i}) = \frac{1}{n}\cdot n\mu = \mu$$

so $$E(\bar{\mathrm{X}}) = \mu$$

The unbiased estimate of mathematical expectation is

$$\hat{\mu} = \bar{x}$$

Example 2. Sample variance

$$s_n^2 = \frac{1}{n}\sum_{i=1}^{n}(x_i - \bar{x})^2$$

Because $$E(S_n^2) = \frac{1}{n}\sum_{i=1}^{n} E\left[(\mathrm{X}_i - \bar{\mathrm{X}})^2\right] = \frac{1}{n}\sum_{i=1}^{n} E(\mathrm{X}_i^2) - E(\bar{\mathrm{X}}^2) = E(\mathrm{X}^2) - E(\bar{\mathrm{X}}^2)$$

where X and X_i have the same distribution.

$$E(\mathrm{X}^2) = \mathrm{Var}(\mathrm{X}) + \left[E(\mathrm{X})\right]^2 = \sigma^2 + \mu^2$$

$$E(\bar{\mathrm{X}}^2) = \mathrm{Var}(\bar{\mathrm{X}}) + \left[E(\bar{\mathrm{X}})\right]^2 = \mathrm{Var}(\bar{\mathrm{X}}) + \mu^2$$

$$\mathrm{Var}(\bar{\mathrm{X}}) = \frac{1}{n^2}\sum_{i=1}^{n}\mathrm{Var}(\mathrm{X}_i) = \frac{\sigma^2}{n}$$

So $$E(X^2) = \sigma^2 + \mu^2$$

$$E(\bar{X}^2) = \frac{\sigma^2}{n} + \mu^2$$

$$E(S_n^2) = E(X^2) - E(\bar{X}^2) = \sigma^2 - \frac{\sigma^2}{n} = \frac{n-1}{n}\sigma^2 \neq \sigma^2$$

Cross-multiplying gives

$$E\left(\frac{n}{n-1}S_n^2\right) = \sigma^2$$

Therefore, the unbiased estimate of the variance is

$$\widehat{\sigma^2} = \frac{n}{n-1}s_n^2 = s_{n-1}^2$$

The biased estimate of the variance is

$$\widehat{\sigma^2} = s_n^2$$

The systematic error of $\widehat{\sigma^2} = s_{n-1}^2$ is

$$E(\varepsilon) = E(S_{n-1}^2) - \sigma^2 = 0$$

The systematic error of $\widehat{\sigma^2} = s_n^2$ is

$$E(\varepsilon) = E(S_n^2) - \sigma^2 = \frac{n-1}{n}\sigma^2 - \sigma^2 = -\frac{\sigma^2}{n} \neq 0$$

It can be seen from here that the meaning of unbiased estimation is that there is no systematic error. That is,

$$E(\varepsilon) = 0$$

6.5. Examples

Example 1. To check the quality of a batch of parts, 10 pieces are taken from the batch and their lengths are measured as follows (Unit: mm).

22.36	22.35	22.33	22.37	22.35
22.34	22.38	22.36	22.35	22.32

(1) In this question, what do population, individual, sample, and sample size refer to?

(2) Calculate the sample mean.

Solution.

(1) Length of each part = individual x_i $(i = 1, 2, \ldots, \mathrm{N})$

{population} = {all lengths of this batch of parts} = $\{x_1, x_2, \ldots, x_\mathrm{N}\} = \{\mathrm{X}\}$

population size = N

{sample} = {22.36, 22.35,…, 22.32}, sample size $n = 10$

(2) sample mean

$$\bar{x} = \frac{1}{n}\sum_{i=1}^{n} x_i = \frac{1}{10}(22.36 + 22.35 + \ldots + 22.32) = 22.351$$

Example 2. A class took a math test with a total score of 7 points. The results are as follows. 7 people scored 7 points, 14 people scored 6 points, 17 people got 5 points, 8 people got 4 points, and 2 people got 2 points. Calculate the average score of the class for this exam.

Solution. Because the whole class score is a population, so the average score is

$$E(\mathrm{X}) = \frac{\sum_{i=1}^{k} x_i f_i}{\sum_{i=1}^{k} f_i} = \frac{7\times7 + 6\times14 + 5\times17 + 4\times8 + 2\times2}{7+14+17+8+2} = \frac{254}{48} \approx 5.29$$

Example 3. Two machines, *A* and *B*, produce one component simultaneously. In 10 days, the number of defective products produced by the two machines every day is

A	0	1	0	2	2	0	3	1	2	4
B	2	3	1	1	0	2	1	1	0	1

Calculate the mean and variance of these two samples. From the calculation results, which machine performs better?

Solution. Sample mean

$$\bar{x}_A = \frac{1}{n}\sum_{i=1}^{n} x_i = \frac{1}{10}(0+1+...+4) = 1.5$$

$$\bar{x}_B = \frac{1}{n}\sum_{i=1}^{n} x_i = \frac{1}{10}(2+3+...+1) = 1.2$$

Sample variance

$$s_A^2 = \frac{1}{n}\sum_{i=1}^{n}(x_i - \bar{x}_A)^2 = \frac{1}{n}\sum_{i=1}^{n} x_i^2 - \bar{x}_A^2 = \frac{1}{10}(0^2 + 1^2 + ... + 4^2) - 1.5^2 = 1.65$$

$$s_B^2 = \frac{1}{n}\sum_{i=1}^{n}(x_i - \bar{x}_B)^2 = \frac{1}{n}\sum_{i=1}^{n} x_i^2 - \bar{x}_B^2 = \frac{1}{10}(2^2 + 3^2 + ... + 1^2) - 1.2^2 = 0.76$$

Since $s_B^2 < s_A^2$, it can be estimated that the performance of machine B is better (that is, the stability is better).

Example 4. A sample of size $n = 36$ is randomly selected from the population N $(52, 6.3^2)$. Find the probability that the sample mean $\bar{X}$ falls between 50.8 and 53.8.

Solution. According to the formula of normal distribution

$$X \sim N\,[E(X), \text{Var}(X)],$$ that is, X follows a normal distribution.

Let $X = \bar{X}$. Substituting gives

$$\bar{X} \sim N\,[E(\bar{X}), \text{Var}(\bar{X})],$$ $\bar{X}$ also follows a normal distribution.

Because

$$E(\bar{X}) = E\left(\frac{1}{n}\sum_{i=1}^{n} X_i\right) = E(X)$$

$$\text{Var}(\bar{X}) = \text{Var}\left(\frac{1}{n}\sum_{i=1}^{n} X_i\right) = \frac{\text{Var}(X)}{n}$$

so $$\bar{X} \sim N\,[E(\bar{X}),\ Var(\bar{X})] = N(\mu,\ \frac{\sigma^2}{n})$$

The derivation of the above formula is an important property of the normal distribution. Hence, we have

$$\text{population} \sim N\,(52,\ 6.3^2)$$

That is, population X follows a normal distribution. So

$$\bar{X} \sim N\,(52,\ \frac{6.3^2}{36})$$

That is, $\bar{X}$ also follows a normal distribution ($n = 36$). Then

$$P\{50.8 < \bar{X} < 53.8\} \approx 0.830$$

Example 5. A sample of size $n = 5$ is randomly selected from the population N (12, 4). Find the probability that the difference between the sample mean and the population mean is greater than 1.

Solution. Because

$$\text{population} \sim N\,(12,\ 4)$$

That is, population X follows a normal distribution. So

$$\bar{X} \sim N\,(12,\ \frac{4}{5})$$

That is, $\bar{X}$ also follows a normal distribution ($n = 5$).

Then $$P\{|\bar{X} - E(X)| > 1\} = P\{|\bar{X} - 12| > 1\} = 1 - P\{|\bar{X} - 12| \le 1\}$$

$$= 1 - P\{11 \le \bar{X} \le 13\} \approx 1 - 0.736 = 0.264$$

That is, $$P\{|\bar{X} - 12| > 1\} \approx 0.264$$

Exercise 6

1. Sugar bags are filled by two machines, A and B. A sample of 7 bags of sugar was taken randomly from the sugar filled by the two machines. Their respective weights in grams are given below.

A	102	101	99	103	98	99	98
B	110	105	90	85	75	115	110

a. Estimate the mean and variance of the weight of sugar bags filled by the two machines.

b. Which machine produces sugar bags with a more stable weight?

2. The results of two long jump athletes, A and B, in the 10 test competitions are as follows (in meters):

A	5.85	5.93	6.07	5.91	5.99	6.13	5.98	6.05	6.00	6.19
B	6.11	6.08	5.83	5.92	5.84	5.81	6.18	6.17	5.85	6.21

Calculate the variance of two samples. Estimate which athlete's performance is more stable based on the calculation results.

3. For a sample, $s_n^2 = 0.63$, $s_{n-1}^2 = 0.7$, find the value of sample size n.

4. Given that s_x is the standard deviation of $x_1, x_2, ..., x_n$, s_y is the standard deviation of $3x_1+5, 3x_2+5, ..., 3x_n+5$, prove that $s_y = 3s_x$.

5. A distribution has a known mean μ and variance σ^2. X_1, X_2, and X_3 are a random sample of three independent observations taken from the population. Given that $Y_1 = \frac{X_1+X_2+X_3}{3}$, $Y_2 = \frac{X_1+2X_2+3X_3}{6}$, $Y_3 = \frac{X_1+X_2+2X_3}{4}$ are three estimators for μ,

a. determine which of them are unbiased estimators for μ.

b. Which of any unbiased estimators has the smallest variance?

Chapter 7 Statistical cases

7.1. Independence test

Example 1. To study whether chronic bronchitis is related to smoking, 339 people over the age of 50 were surveyed. The survey results are shown in the table below.

	chronic bronchitis	No chronic bronchitis	total
smoking	43	162	205
no smoking	13	121	134
total	56	283	339

Is there any relationship between chronic bronchitis and smoking habits among people over 50 years old?

Analysis: The table given in Example 1 is called a 2 × 2 contingency table. Now, we need to consider two conditions for people over 50 years old: whether they smoke and whether they suffer from chronic bronchitis. Based on the four data arranged in two rows and two columns in the table, we hope to test whether there is a relationship between the two conditions. This test is called the independence test of the 2 × 2 contingency table.

To discuss the problem clearly and facilitate the generalisation to the general situation, we use letters to replace the events and data in the 2 × 2 contingency table and obtain a 2 × 2 contingency table represented by letters, as shown in the table below.

	chronic bronchitis (B)	No chronic bronchitis (B')	total
smoking (A)	n_{11}	n_{12}	n_{1+}
no smoking (A')	n_{21}	n_{22}	n_{2+}
total	n_{+1}	n_{+2}	n

In the table: $n_{+1} = n_{11} + n_{21}$, $n_{+2} = n_{12} + n_{22}$, $n_{1+} = n_{11} + n_{12}$, $n_{2+} = n_{21} + n_{22}$,

$n = n_{11} + n_{21} + n_{12} + n_{22}$

First, when there is a relationship between smoking (A) and chronic bronchitis (B), that is, A and B are not independent, we can infer using the probability method that

$$P(AB) = P(B|A)P(A)$$

Secondly, when there is no relationship between smoking (A) and suffering from chronic bronchitis (B), that is, A and B are independent, using the probability method to infer, we get

$$P(AB) = P(A)P(B)$$

We use the expression H_0 to represent the above equation. That is,

$$H_0: \quad P(AB) = P(A)P(B)$$

H_0 is called a statistical hypothesis. When H_0 is true, the following three equations are also true.

$$P(A'B) = P(A')P(B)$$
$$P(AB') = P(A)P(B')$$
$$P(A'B') = P(A')P(B')$$

According to the statistical definition of probability, the probabilities of many events mentioned above can be estimated using the corresponding relative frequencies $f(\bullet)$.

Because $$\lim_{n\to\infty} f(\bullet) = \lim_{n\to\infty} \frac{n(\bullet)}{n} = P(\bullet)$$

When n is relatively large, we have

$$f(\bullet) = \frac{n(\bullet)}{n} \approx P(\bullet)$$

Hence $$\frac{n_{11}}{n} \approx P(AB) \qquad \frac{n_{1+}}{n} \approx P(A) \qquad \frac{n_{+1}}{n} \approx P(B) \quad \ldots$$

Let

$$\chi^2 = \frac{\left(\frac{n_{11}}{n} - \frac{n_{1+}}{n} \cdot \frac{n_{+1}}{n}\right)^2}{\frac{n_{1+}}{n} \cdot \frac{n_{+1}}{n}} + \frac{\left(\frac{n_{12}}{n} - \frac{n_{1+}}{n} \cdot \frac{n_{+2}}{n}\right)^2}{\frac{n_{1+}}{n} \cdot \frac{n_{+2}}{n}} + \frac{\left(\frac{n_{21}}{n} - \frac{n_{2+}}{n} \cdot \frac{n_{+1}}{n}\right)^2}{\frac{n_{2+}}{n} \cdot \frac{n_{+1}}{n}} + \frac{\left(\frac{n_{22}}{n} - \frac{n_{2+}}{n} \cdot \frac{n_{+2}}{n}\right)^2}{\frac{n_{2+}}{n} \cdot \frac{n_{+2}}{n}}$$

$$= \frac{n(n_{11}n_{22} - n_{12}n_{21})^2}{n_{1+}n_{2+}n_{+1}n_{+2}}$$

That is, $$\chi^2 = \frac{n(n_{11}n_{22} - n_{12}n_{21})^2}{n_{1+}n_{2+}n_{+1}n_{+2}}$$

This is the statistic expression of χ^2 (pronounced chi-square).

In the expression of χ^2, the properties of the numerator of the first fractional term are

- when A and B are related, that is, A and B are not independent of each other

  $$\frac{n_{11}}{n} - \frac{n_{1+}}{n} \cdot \frac{n_{+1}}{n} \approx \mathrm{P(AB) - P(A)P(B)} \neq \mathrm{P(AB) - P(B|A)P(A)} = 0$$

  That is, $\frac{n_{11}}{n} - \frac{n_{1+}}{n} \cdot \frac{n_{+1}}{n} \approx a \qquad 0 < |a| < 1$

- when A and B have no relationship, that is, A and B are independent of each other

  $$\frac{n_{11}}{n} - \frac{n_{1+}}{n} \cdot \frac{n_{+1}}{n} \approx \mathrm{P(AB) - P(A)P(B)} = \mathrm{P(AB) - P(B|A)P(A)} = 0$$

  That is, $\frac{n_{11}}{n} - \frac{n_{1+}}{n} \cdot \frac{n_{+1}}{n} \approx 0$

In the expression of χ^2, because the numerators of the four fractions all have the same properties as above, so

χ^2 for (A, B) is related > χ^2 for (A, B) is not related

Then there should be a critical value as a criterion for judging the relationship between A and B. After studying the distribution of statistics χ^2, there are indeed two critical values: 3.841 and 6.635.

When calculation based on specific data gives $\chi^2 > 3.841$, the probability $P\{\chi^2 > 3.841\} = 0.95$. That is, there is 95% confidence that events A and B are related. That is, A and B are not independent of each other. The probability of being wrong is 5%.

When $\chi^2 > 6.635$, the probability $P\{\chi^2 > 6.635\} = 0.99$. That is, there is 99% certainty that events A and B are related. That is, A and B are not independent of each other. The probability of being wrong is 1%.

When $\chi^2 \le 3.841$, we consider that events A and B have no relationship. That is, A and B are mutually independent.

Solution to Example 1. From the formula, we have

$$\chi^2 = \frac{339(43\times121-162\times13)^2}{205\times134\times56\times283} \approx 7.469$$

Because 7.469 > 6.635, therefore, we are 99% confident to say that chronic bronchitis in people over 50 years old is related to smoking habits. That is, A and B are not independent of each other.

Example 2. A survey on airsickness among male and female passengers during a flight in adverse weather conditions is shown in the table below. Based on this information, do you think men are more likely to get airsick than women when flying in bad weather?

	airsickness	no airsickness	total
male	24	31	55
female	8	26	34
total	32	57	89

Solution. From the formula, we have

$$\chi^2 = \frac{89(24\times 26-31\times 8)^2}{55\times 34\times 32\times 57} \approx 3.689$$

Because 3.689 < 3.841, we can say that airsickness has no relationship with male and female gender. That is, A and B are independent of each other. Although the proportion of men ($\frac{24}{55}$) suffering from airsickness on this flight is higher than that of women ($\frac{8}{34}$), we cannot assume that men are more prone to airsickness than women during flights in bad weather.

Note:

When using the statistic χ^2 to test the independence of a 2 × 2 contingency table, all four data in the table must be greater than 5. This must be considered when selecting the sample size. The four data in this example, 24, 31, 8, and 26, are all greater than 5, which meets this requirement.

Example 3. Snoring affects others' rest and may also be related to certain diseases. The table below shows data from a survey. Is there a connection between snoring every night and heart disease?

	Have heart disease	No heart disease	total
snore every night	30	224	254
No snoring	24	1355	1379
total	54	1579	1633

Solution. From the formula, we have

$$\chi^2 = \frac{1633(30\times1355-224\times24)^2}{254\times1379\times54\times1579} \approx 68.033$$

Because 68.33 > 6.635, so we are 99% sure that snoring every night is related to heart disease. That is, A and B are not independent of each other.

Note:

This example is similar to Example 1. When we say "there is a relationship between snoring every night and heart disease" or "there is a relationship between chronic bronchitis and smoking habits", we are referring to statistical relationships. Do not mistake them for causal relationships. Specifically speaking, every person who snores every night cannot be said to have a heart disease. In fact, we can also see from the 2 × 2 contingency table that his probability of suffering from heart disease is only $\frac{30}{254}$, slightly more than $\frac{1}{10}$. Whether he has heart disease should be determined by medical examination. It is no longer a matter of statistics. □

This section discusses whether two events are independent through three examples. In the independence test of the 2 × 2 contingency table, we choose the statistic χ^2, and its value can be used to infer whether the independence is true. Independence tests are widely used in the disciplines of biostatistics and medical statistics. Independence tests are also often used when dealing with data obtained from social issues.

Exercise 7.1

1. A survey of the colourblindness of a large number of adults gives the following information.

	colourblind	not colourblind
female	10	800
male	55	945

Determine whether or not the gender of a person plays a part in whether or not they are colourblind.

2. A survey asked adults whether they read nutrition facts labels when buying food. The data obtained is as follows.

	Read the nutrition facts label	do not read the nutrition facts label
male	32	10
female	20	26

Is there a relationship between a person's gender and whether they read nutrition facts labels?

3. To understand the effectiveness of a particular influenza vaccine in preventing influenza, the data obtained from a survey are as follows.

	did not have an influenza	had an influenza
Got the influenza vaccine	268	255
did not get the influenza vaccine	232	294

Is the influenza vaccine effective in preventing influenza?

4. A factory conducted an investigation to find out whether there was a difference in product qualification rates between the day and night shifts and obtained the following data.

	qualified products	defective products
day shift	242	22
night shift	63	20

Is there any difference in product qualification rates between the day and night shifts?

5. To determine the effectiveness of a specific fertiliser, 1,000 plants were tested. Of the 200 plants that were not fertilised, 103 grew well. Of the 800 plants that were fertilised, 678 were growing well. Is the effect of fertilisation significant?

7.2. Regression analysis

7.2.1. Correlation coefficient of random variables X and Y

Consider using a linear function of X, $a\mathrm{X} + b$, to approximately represent Y. We use the mean square error e to measure the strength of the approximation of Y by $a\mathrm{X} + b$.

$$e = E\{[\mathrm{Y} - (a\mathrm{X} + b)]^2\} \qquad (1)$$

$$= E(\mathrm{Y}^2) + a^2E(\mathrm{X}^2) + b^2 - 2aE(\mathrm{XY}) + 2abE(\mathrm{X}) - 2bE(\mathrm{Y})$$

The smaller the value of e, the better the approximation between $a\mathrm{X} + b$ and Y. Therefore, we will take the values of a and b so that e is minimised.

In the following, we will find the values of a and b in the best approximation expression $a\mathrm{X} + b$. To achieve this, take the partial derivatives of e with respect to a and b, respectively, set them equal to 0, and we get

$$\frac{\partial e}{\partial a} = 2aE(\mathrm{X}^2) - 2E(\mathrm{XY}) + 2bE(\mathrm{X}) = 0$$

$$\frac{\partial e}{\partial b} = 2b + 2aE(\mathrm{X}) - 2E(\mathrm{Y}) = 0$$

Solving simultaneously gives

$$a = \frac{\mathrm{Cov(X,Y)}}{\mathrm{Var(X)}}$$

$$b = E(\mathrm{Y}) - E(\mathrm{X})\frac{\mathrm{Cov(X,Y)}}{\mathrm{Var(X)}}$$

where $\mathrm{Cov(X, Y)} = E\{[\mathrm{X} - E(\mathrm{X})][\mathrm{Y} - E(\mathrm{Y})]\}$. Cov(X, Y) is called covariance.

Substituting the values of a and b into equation (1) and rearranging them gives

$$E\{[\mathrm{Y} - (a\mathrm{X} + b)]^2\}_{\min} = (1 - \rho^2)\mathrm{Var(Y)} \tag{2}$$

where $\rho = \dfrac{\mathrm{Cov(X,Y)}}{\sqrt{\mathrm{Var(X)}}\sqrt{\mathrm{Var(Y)}}}$. ρ is called the correlation coefficient between X and Y.

Sample correlation coefficient

For a random sample, we can use the sample statistic instead.

sample correlation coefficient

$$\mathrm{r} = \hat{\rho} = \frac{\widehat{Cov(X,Y)}}{\sqrt{\widehat{Var(X)}}\sqrt{\widehat{Var(Y)}}}$$

$$= \frac{\frac{1}{n}\sum_{i=1}^{n}(x_i - \bar{x})(y_i - \bar{y})}{\sqrt{\frac{1}{n}\sum_{i=1}^{n}(x_i - \bar{x})^2}\sqrt{\frac{1}{n}\sum_{i=1}^{n}(y_i - \bar{y})^2}} = \frac{\sum_{i=1}^{n}(x_i - \bar{x})(y_i - \bar{y})}{\sqrt{\sum_{i=1}^{n}(x_i - \bar{x})^2}\sqrt{\sum_{i=1}^{n}(y_i - \bar{y})^2}}$$

For the convenience of calculation, we introduce the following symbols.

$$\mathrm{S_{XX}} = \sum_{i=1}^{n}(x_i - \bar{x})^2$$

$$S_{yy} = \sum_{i=1}^{n}(y_i - \bar{y})^2$$

$$S_{xy} = \sum_{i=1}^{n}(x_i - \bar{x})(y_i - \bar{y})$$

In this way, r can be written as

sample correlation coefficient

$$r = \frac{S_{xy}}{\sqrt{S_{xx}}\sqrt{S_{yy}}}$$

Properties of the population correlation coefficient ρ

Because $E\{[Y - (aX + b)]^2\} \geq 0$ Var(Y) > 0 ($Y \neq$ constant)

from equation (2), we can get

$$(1 - \rho^2) \geq 0$$

so $0 \leq |\rho| \leq 1$

or $-1 \leq \rho \leq 1$

Because the mean square error e is a function of $|\rho|$. When $|\rho|$ is larger, e is smaller, and we usually say that the strength of the linear correlation between X and Y is higher.

Especially when $|\rho| = 1$, $e = 0$, X and Y have perfect linear correlation.

When $|\rho|$ is smaller, e is larger, and we say that the strength of the linear correlation between X and Y is relatively weaker.

Especially when $|\rho| = 0$, $e = \text{Var}(Y)$ is the maximum value, and we say that X and Y are not linearly correlated.

When $0 < \rho \leq 1$, positive linear correlation;

When $-1 \leq \rho < 0$, negative linear correlation.

Represent with absolute value.

When $0 < |\rho| \leq 1$, linear correlation;

When $\rho = 0$, no linear correlation.

Properties of sample correlation coefficient r

$|r| \leq 1$. The closer $|r|$ is to 1, the stronger the strength of the linear correlation between x and y; The smaller $|r|$ is, the weaker the strength of the linear correlation between x and y.

After studying the distribution of r statistics, we can get

$r_{0.05} < |r| \leq 1$ linear correlation

$0 \leq |r| \leq r_{0.05}$ no linear correlation

where $r_{0.05}$ is called the critical value. $0.05 \approx 0$ represents the probability of making judgement mistake.

For example, when $3 \leq n \leq 30$ (n pairs of data)

$0.361 \leq r_{0.05} \leq 0.997$ The larger n is, the smaller $r_{0.05}$.

The method of finding $r_{0.05}$ is omitted here. The value of $r_{0.05}$ can be found in the special mathematical table – Pearson correlation coefficient critical value table where $N = n$. For example, $n = 10 \Leftrightarrow n - 2 = 8$.

7.2.2. Unary linear regression

Let there be some correlation between random variables y and x. Here, x is a variable that can be controlled or observed accurately, such as age, temperature in the experiment, applied pressure, voltage and time, etc. In other words, we can determine n values, $x_1, x_2, ..., x_n$. That is, we treat x as an ordinary variable. We only discuss this case here.

Since y is a random variable, for every specific value of x, y has its distribution. If the mathematical expectation of y exists, its value depends on the value of x. That is, the mathematical expectation of y is a function of x, denoted as

$$E(\mathrm{Y}|\mathrm{X}) = \mu_{y|x} \quad \text{or} \quad \mu(x)$$

$\mu(x)$ is called the regression of y on x.

Since the magnitude of $\mu(x)$ reflects to some extent the magnitude of the observed value of random variable y at x, when x takes a specific value, it can estimate the value of the random variable y.

We take a set of non-identical values for x, $x_1, x_2, ..., x_n$, and conduct independent experiments to obtain n pairs of observed values.

$$(x_1, y_1), (x_2, y_2), ..., (x_n, y_n)$$

where y_i is the observed result of random variable y at $x = x_i$.

These n pairs of observations are a sample of size n. The first problem we must solve is how to use the sample to estimate $\mu(x)$, the regression of y on x. To do this, we must guess the functional form of $\mu(x)$. In some problems, we can directly know the functional form of $\mu(x)$. Otherwise, we can plot the corresponding point of each pair of observations (x_i, y_i) in the Cartesian coordinate plane (as Graph 7.1 in the example below). This type of diagram is called a scatter diagram. A scatter diagram can help us roughly to observe the functional form of $\mu(x)$.

Example. At a junior tournament, 10 young athletes participated in the long jump competition. The table below records each athlete's age and jumping distance.

Athlete	1	2	3	4	5	6	7	8	9	10
Age (years)	10	16	13	11	17	12	18	14	15	16
jumping distance(m)	4.3	7.0	5.8	5.1	7.4	5.8	7.6	5.9	6.6	7.6

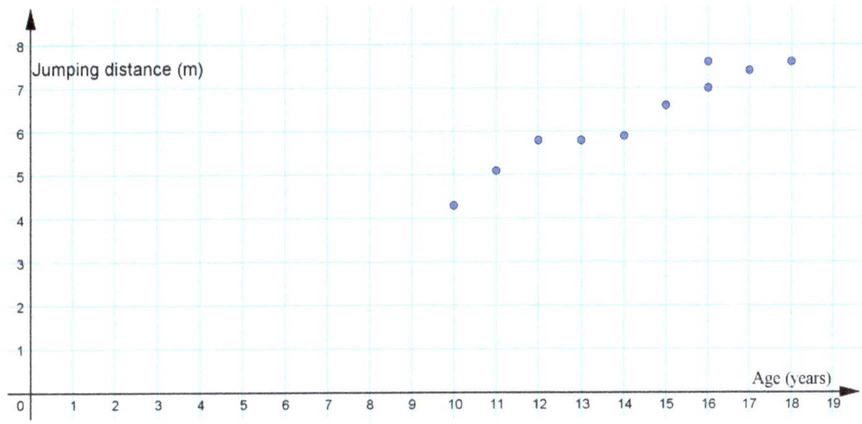

Graph 7.1 Scatter diagram of data above

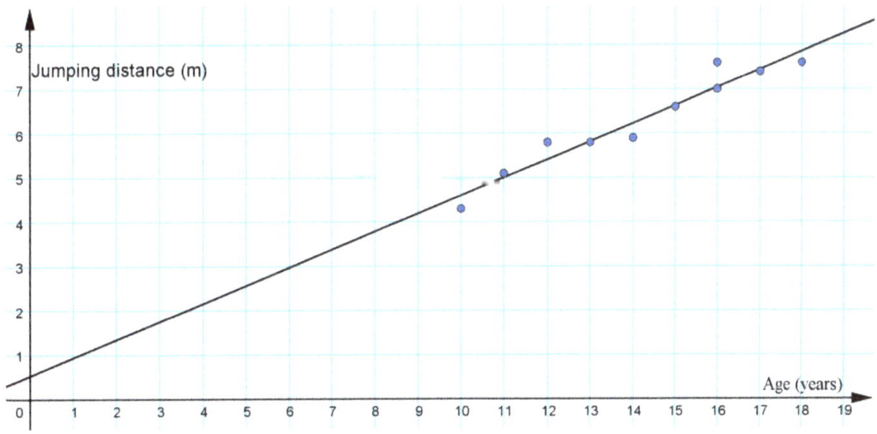

Graph 7.2 Scatter diagram of data above with the line of best fit

Here, the independent variable x (age) is an ordinary variable, and y (jumping distance) is a random variable. It can be roughly seen from Graph 7.2 that $\mu(x)$ has the form of a linear function $ax + b$.

Let the regression of y on x be $\mu(x)$. The problems of using samples to estimate $\mu(x)$ are called problems of finding the regression of y on x. In particular, if $\mu(x)$ is a linear function, $\mu(x) = ax + b$, the problem of estimating $\mu(x)$ is called a unary linear regression problem. Here we only discuss this problem.

Let $\quad y \sim N(ax + b, \sigma^2)$

Because $\quad y - \mu(x) = \varepsilon \text{ (error)} \sim N(0, \sigma^2)$

so $\quad y = \mu(x) + \varepsilon = ax + b + \varepsilon$

That is, $\quad y = ax + b + \varepsilon \qquad (3)$

where a, b are unknown parameters.

We call the equation (3) the model of unary linear regression.

If the estimates of a and b, $\hat{a}$ and $\hat{b}$, are obtained from the sample, then for a given x, we have an estimate of $\mu(x)$.

$$\hat{\mu}(x) = \hat{a}x + \hat{b}$$

Hence

$$\hat{y} = \hat{\mu}(x) + \hat{\varepsilon} = \hat{a}x + \hat{b} + \hat{\varepsilon}$$

If x is linearly correlated with y, that is, $r_{0.05} < |r| \le 1$, the estimate of error $\hat{\varepsilon} \approx 0$ (or $\hat{\varepsilon} = 0$), then we have

$$\hat{y} \approx \hat{\mu}(x) = \hat{a}x + \hat{b}$$

It can be approximately treated as

$$\hat{y} = \hat{a}x + \hat{b}$$

This equation can be used as an estimate of y or $\mu(x) = ax + b$. We call it the linear regression equation or regression equation of y on x.

It can be seen from the above that the estimated value $\hat{y}$ is on the regression line, but the value of y is not necessarily on the regression line. That is,

$$y = \hat{y} = \hat{a}x + \hat{b} \qquad \text{or} \qquad y \approx \hat{y} = \hat{a}x + \hat{b}$$

(The two equations show that y is linearly correlated to x.)

If the sample correlation coefficient $|r| > r_{0.05}$, then the regression equation

$$\hat{y} = \hat{a}x + \hat{b} \qquad \text{It is linearly correlated.}$$

If $|r| \leq r_{0.05}$, then the regression equation

$$\hat{y} \neq \hat{a}x + \hat{b} \qquad \text{It is linearly uncorrelated.}$$

It is meaningless to find the regression equation at this time.

Maximum likelihood estimation method

(1) The population X is discrete

If $x_1, x_2, ..., x_n$ is a sample value of the population X, and the corresponding probability distribution is

$$P\{X = x_i\} = p(x_i, \theta) \qquad i = 1, 2, \dots, n$$

then, the probability that the observed values $x_1, x_2, ..., x_n$ occur simultaneously is

$$L(\theta) = \prod_{i=1}^{n} p(x_i, \theta)$$

where θ is the parameter to be estimated;

$L(\theta)$ is the probability of simultaneous occurrence, and we call it the likelihood function of the sample;

$\prod_{i=1}^{n}$ is a product operator.

(2) The population X is continuous

The form of probability density $f(x)$ is known, and its probability distribution is

$$P\{X = x_i\} = f(x_i, \theta) \lim_{\Delta x_i \to 0} \Delta x_i \qquad i = 1, 2, \ldots, n$$

The probability of getting all observed values $x_1, x_2, \ldots, x_n$ occurring simultaneously is

$$L(\theta) = \prod_{i=1}^{n} f(x_i, \theta) \lim_{\Delta x_i \to 0} \Delta x_i$$

In many cases, $L(\theta)$ has a maximum value. $p(x_i, \theta)$ and $f(x_i, \theta)$ are differentiable with respect to θ. Then, θ can be found from the equation $\frac{dL(\theta)}{d\theta} = 0$.

In addition, since

$$\frac{d \ln L(\theta)}{d\theta} = \frac{d \ln L(\theta)}{dL(\theta)} \cdot \frac{dL(\theta)}{d\theta} = \frac{1}{L(\theta)} \cdot \frac{dL(\theta)}{d\theta}$$

so

$$\frac{dL(\theta)}{d\theta} = L(\theta) \frac{d \ln L(\theta)}{d\theta}$$

When $\frac{dL(\theta)}{d\theta} = 0$, we have $\frac{d \ln L(\theta)}{d\theta} = 0$. The θ solved from the two equations is the same. Therefore $\hat{\theta}$, the maximum likelihood estimate of θ, can be obtained from the equation

$$\frac{d \ln L(\theta)}{d\theta} = 0$$

From this equation, we can see that the constant derivative is

$$(\ln\prod_{i=1}^{n}\lim_{\Delta x_i\to 0}\Delta x_i)' = 0$$

Hence, we only need to consider the function after the omission of the factor

$\prod_{i=1}^{n}\lim_{\Delta x_i\to 0}\Delta x_i$.

That is, $$L(\theta)=\prod_{i=1}^{n}f(x_i,\theta)\lim_{\Delta x_i\to 0}\Delta x_i=\prod_{i=1}^{n}f(x_i,\theta)\cdot\prod_{i=1}^{n}\lim_{\Delta x_i\to 0}\Delta x_i$$

After omission, it becomes

$$L(\theta) = \prod_{i=1}^{n}f(x_i,\theta)$$

↑

(This is called joint probability density)

- Using the formula $\frac{d\ln L(\theta)}{d\theta} = 0$, we can solve for the values of $\hat{a}$ and $\hat{b}$ in the regression equation.

Linear model $y \sim \mathrm{N}\,(ax + b, \sigma^2)$

The probability density of the normal distribution is

$$f(y)=\frac{1}{\sigma\sqrt{2\pi}}e^{-\frac{[y-\mu(x)]^2}{\sigma^2}}$$

The likelihood function is

$$L=\prod_{i=1}^{n}f(y_i)=\left(\frac{1}{\sigma\sqrt{2\pi}}\right)^n\cdot e^{-\frac{1}{\sigma^2}\sum_{i=1}^{n}[y_i-\mu(x_i)]^2}$$

Taking logarithm gives

$$\ln L = \ln\left(\frac{1}{\sigma\sqrt{2\pi}}\right)^n-\frac{1}{\sigma^2}\sum_{i=1}^{n}\left[y_i-\mu(x_i)\right]^2\ln e$$

Let $\varepsilon = \sum_{i=1}^{n}\left[y_i - \mu(x_i)\right]^2$

Since L has a maximum, so ε has a minimum.

Let $\frac{\partial \ln L}{\partial a} = 0 \qquad \frac{\partial \ln L}{\partial b} = 0$

Because $\frac{\partial \ln L}{\partial a} = [\ln\left(\frac{1}{\sigma\sqrt{2\pi}}\right)^n]' - \frac{1}{\sigma^2}\frac{\partial \varepsilon}{\partial a} = -\frac{1}{\sigma^2}\frac{\partial \varepsilon}{\partial a}$

so $\frac{\partial \ln L}{\partial a} = 0$ gives $\frac{\partial \varepsilon}{\partial a} = 0$

Similarly $\frac{\partial \ln L}{\partial b} = 0$ gives $\frac{\partial \varepsilon}{\partial b} = 0$

Therefore $\frac{\partial \varepsilon}{\partial a} = \frac{\partial}{\partial a}\sum_{i=1}^{n}[y_i - (ax_i + b)]^2 = 0$

$$\frac{\partial \varepsilon}{\partial b} = \frac{\partial}{\partial b}\sum_{i=1}^{n}[y_i - (ax_i + b)]^2 = 0$$

Solving these two equations simultaneously can give the values of $\hat{a}$ and $\hat{b}$.

The solution process is omitted, and we only give the solution results.

$$\hat{a} = \frac{\sum_{i=1}^{n}(x_i - \overline{x})(y_i - \overline{y})}{\sum_{i=1}^{n}(x_i - \overline{x})^2}$$

$$\hat{b} = \overline{y} - \hat{a}\overline{x}$$

If we substitute $\hat{b} = \overline{y} - \hat{a}\overline{x}$ into the regression equation

$$\hat{y} = \hat{a}x + \hat{b}$$

then the regression equation becomes

$$\hat{y} = \overline{y} + \hat{a}(x - \overline{x})$$

This equation shows that when $x = \overline{x}$, $\hat{y} = \overline{y}$.

The regression line passes through the scatter diagram's geometric centre, $(\overline{x}, \overline{y})$.

7.2.3. Interpolation and extrapolation

First, consider

- the model of unary linear regression

Because $y \sim N(ax + b, \sigma^2)$

so $y - \mu(x) = \varepsilon \sim N(0, \sigma^2)$

$$y = \mu(x) + \varepsilon = ax + b + \varepsilon$$

$$E(Y) = aE(X) + b + E(\varepsilon)$$

Since $E(\varepsilon) = 0$

so $E(Y) = aE(X) + b$

According to the definition of normal distribution

$$E(Y) = \mu(x)$$

we get $aE(X) + b = ax + b$

When $x = E(X)$, the equality holds true;

When x takes values near $E(X)$, $x \approx E(X)$, the equality approximately holds.

- estimate definition of a regression equation

From the geometric centre $(\bar{x}, \bar{y})$, we have

$$\bar{y} = \hat{a}\bar{x} + \hat{b}$$

According to the definition of an estimate

$$\hat{\mu}(x) = \bar{y}$$

so $\hat{a}x + \hat{b} = \hat{a}\bar{x} + \hat{b}$

When $x = \bar{x}$, the equality holds;

When $x \approx \bar{x}$, the equality approximately holds.

From the above, we can see that if the defining equations of the linear model or regression equation hold, they restrict that x can only take one value, $x = E(X)$ or $x = \bar{x}$. However, in practical applications, to solve problems of linear correlation, we need to relax the value restriction of x. After liberalisation, there are two intervals in which x takes values. One is called the interpolation interval. The other is called the extrapolation interval. The corresponding methods are called the interpolation method and extrapolation method, respectively.

(1) Interpolation

interpolation interval (sample interval) $\qquad x_{\min} \le x \le x_{\max}$

$$\bar{x} = \frac{1}{n}\sum_{i=1}^{n} x_i \begin{pmatrix} = \\ or \\ \approx \end{pmatrix} \frac{x_{\min} + x_{\max}}{2} = \text{(value of the midpoint of the interval)}$$

(This relationship only applies to ordinary variables.)

where $x_{\min}$, $x_{\max}$ are the smallest and biggest data in the sample, respectively.

If x takes values from the interpolation interval, its relative error with $\bar{x}$ is not too large, $\frac{|x-\bar{x}|}{\bar{x}} < 0.2$ (0.2 is an empirical value).

relative error $\qquad \frac{|x-\bar{x}|}{\bar{x}} < 0.2 \quad \Leftrightarrow \quad 0.8\bar{x} < x < 1.2\bar{x}$

Since x takes values near $\bar{x}$ so $\hat{y} = \hat{\mu}(x)$ satisfies approximately the definition of estimate. Then, it is appropriate for $\hat{y} = \hat{\mu}(x)$ to be an estimate of y or $\mu(x)$.

(2) Extrapolation

Extrapolation interval (non-sample interval) $\quad x < x_{\min}$, $\; x > x_{\max}$

If x takes values from the extrapolation intervals, its relative error with $\bar{x}$ is relatively large, $\frac{|x-\bar{x}|}{\bar{x}} \geq 0.2$

$$\text{relative error} \quad \frac{|x-\bar{x}|}{\bar{x}} \geq 0.2 \quad \Leftrightarrow \quad x \leq 0.8\bar{x} \text{ or } x \geq 1.2\bar{x}$$

Since x does not take values near $\bar{x}$, so $\hat{y} = \hat{\mu}(x)$ does not satisfy the definition of estimate. Then, it is inappropriate for $\hat{y} = \hat{\mu}(x)$ to be an estimate of y or $\mu(x)$.

In particular, when $n \geq 10$, $0.95 \leq |r| \leq 1$, y and x are very strongly linearly correlated. That is, $y = \hat{y}$ or $y \rightarrow \hat{y}$. At this time, $\hat{y} = \hat{\mu}(x)$ can be an estimate of y or $\mu(x)$ without restriction.

7.2.4. Examples

Example 1. To understand the relationship between the mother's height and the daughter's height in a certain place, the heights of 9 pairs of mothers and daughters were measured as follows:

mother's height (x cm)	169	170	170	175	168	165	179	163	159
daughter's height (y cm)	168	168	170	176	170	166	177	162	158

a. Calculate $\bar{x}$ and y, correct to two decimal places.

b. Draw a scatter diagram for the data. Locate the mean point $(\bar{x}, \bar{y})$ on the scatter diagram. Also, draw the line of best fit.

c. Find the equation of the regression line of y on x. Calculate the correlation coefficient r.

d. Describe the relationship between the variables.

e. Estimate the height of the daughter of a mother with a height of 171 cm, correct to two decimal places.

f. Estimate the height of the daughter of a mother with a height of 154 cm, correct to two decimal places.

g. Comment on whether the estimates in e. and f. are reliable.

Solution.

a. $$\overline{x} = \frac{169+170+170+175+168+165+179+163+159}{9} = \frac{506}{3} \approx 168.67$$

$$\overline{y} = \frac{168+168+170+176+170+166+177+162+158}{9} = \frac{505}{3} \approx 168.33$$

b.

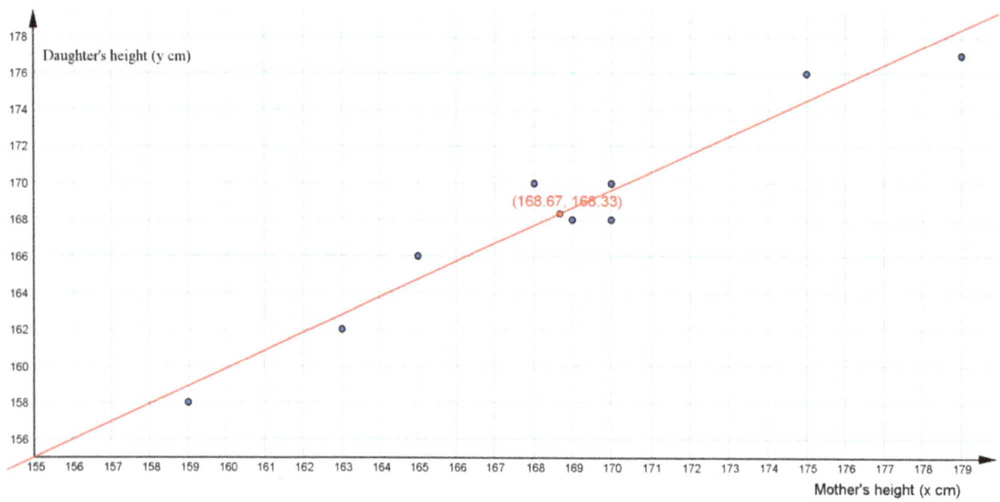

Graph 7.3

c. Use a calculator we get the equation of the regression line of y on x is

$$\hat{y} = 0.97586206x + 3.73793103$$

The correlation coefficient r is

r = 0.97251433 strong linear correlation

d. The variables have a very strong positive linear correlation.

e. For a mother with a height of 171 cm, use the equation of the regression line of y on x, and we can estimate her daughter's height.

$$\hat{y} = 0.97586206 \times 171 + 3.73793103 \approx 170.61$$

f. For a mother with a height of 154 cm, use the equation of the regression line of y on x, and we can estimate her daughter's height.

$$\hat{y} = 0.97586206 \times 154 + 3.73793103 \approx 154.02$$

g. The estimate in e. is an interpolation with a strong linear correlation between the variables. We therefore expect this estimate to be reliable. The estimate in f. is an extrapolation, so the estimate may not be reliable.

Example 2. A set of bivariate data is given in the following table.

x	1	3	4	6	8	9	11	14
y	1	2	4	4	5	7	8	9

a. Find the equation of the regression line of y on x. Calculate the correlation coefficient r.
b. Estimate the value of y when $x = 10$, correct to two decimal places.
c. Given that $y = 6$, estimate the value of x.

Solution.

a. Use a calculator we get the equation of the regression line of y on x is

$$\hat{y} = 0.6363636x + 0.5454545$$

The correlation coefficient r is

r = 0.97700842 strong linear correlation

b. When $x = 10$,

$$\hat{y} = 0.6363636 \times 10 + 0.5454545 \approx 6.91$$

c. We first need to find the equation for the regression line of x on y. Use the calculator we get

$$\hat{x} = 1.5y - 0.5$$

Using this equation and substituting $y = 6$, we get

$$\hat{x} = 1.5 \times 6 - 0.5 = 8.5$$

Example 3. Consider the following bivariate data set when a, b $\in Z^+$.

x	3	5	7	7	9	11
y	8	a	13	b	17	22

The regression of y on x has the equation $\hat{y} = 1.75x + 1.75$. The regression line passes through the mean point $(\overline{x}, \overline{y})$.

a. Given that $\overline{x} = 7$, verify that $\overline{y} = 14$.

b. Given that b – a = 4, find the values of a and b.

Solution.

a. The regression line passes through the mean point $(\overline{x}, \overline{y})$, so substitute $\overline{x} = 7$ into the equation of the regress line, and we get

$$\overline{y} = 1.75 \times 7 + 1.75 = 14$$

b. From $\overline{y} = \dfrac{8 + a + 13 + b + 17 + 22}{6} = 14$ we get

$$a + b = 24$$

And $$b - a = 4$$

Solving the two equations simultaneously gives

$$a = 10, \qquad b = 14$$

Example 4. The following table shows information from a survey on the price of old cars in a country in 2019.

Year of uses (x)	1	2	3	4	5	6	7	8	9	10
Average price (y euro)	30762	21533	18457	12613	8877	6243	5616	3365	2622	2367

x denotes the years the car has been used and y denotes the corresponding average price. Find the regression equation of y on x.

Solution. First, make a scatter plot, as shown in Graph 7.4.

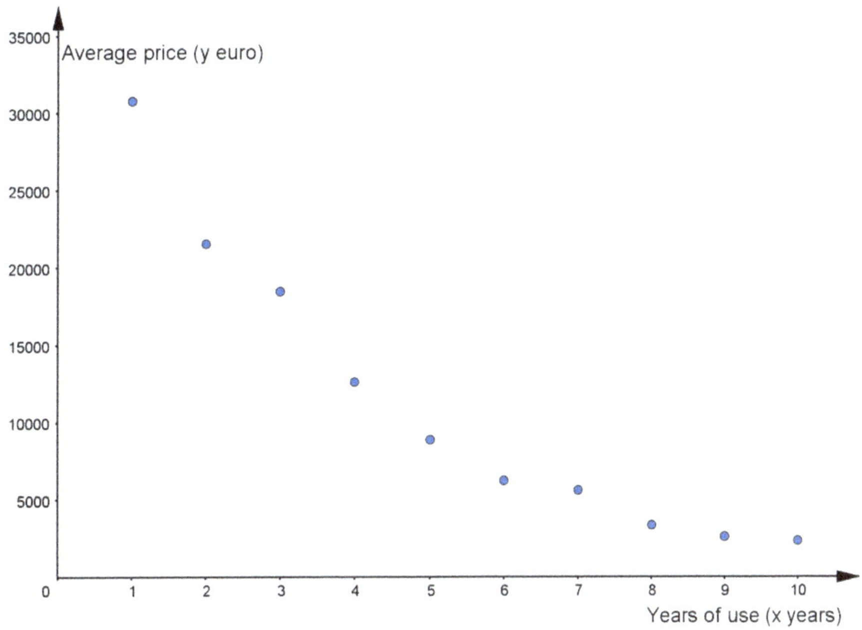

Graph 7.4

It seems that y and x have an exponential relationship, so let $z = \ln y$, make a scatter plot of (x_i, z_i), as shown in Graph 7.5.

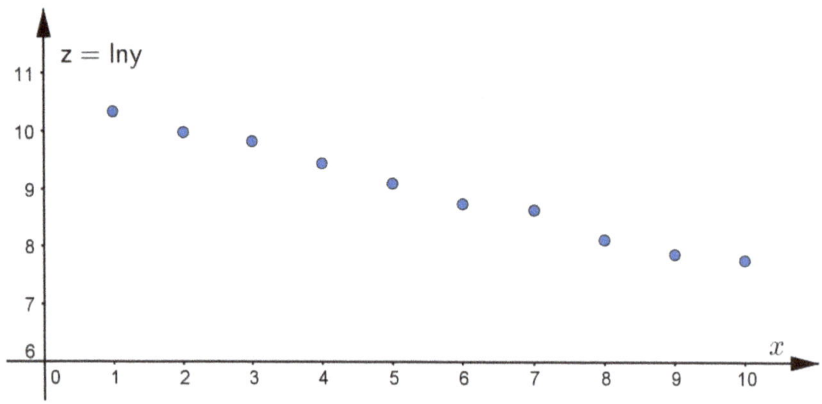

Graph 7.5

It can be seen that all points are basically on a straight line, so let

$$z = ax + b + \varepsilon \qquad \varepsilon \sim N(0, \sigma^2)$$

Using the data (x_i, z_i) and a calculator, we can get

$$|r| = 0.9954969 > 0.632 = r_{0.05} \qquad \text{That is,} \qquad |r| > r_{0.05} \quad \text{linear correlation}$$

We also get

$$\hat{a} = -0.2976415 \qquad \hat{b} = 10.6173411$$

Therefore we have

$$\hat{z} = -0.2976415x + 10.6173411$$

Since $\qquad \ln y = z \qquad$ so $\qquad y = e^z = e^{ax+bx+\varepsilon}$

Hence $\qquad \hat{y} = e^{\hat{z}} = e^{-0.2976415x+10.6173411} \approx 40836.89e^{-0.2976415x}$

That is, the regression equation for the curve is

$$\hat{y} \approx 40836.89e^{-0.2976415x}$$

Example 5. In colour development, it is known empirically that the optical density of the formed dye, Y, and the optical density of the precipitated silver, x, are given by the equation

$$y = Ce^{\frac{a}{x}} \qquad (a < 0)$$

The experimental data currently measured are as follows:

x_i	0.05	0.06	0.25	0.31	0.07	0.10	0.38	0.43	0.14	0.20	0.47
y_i	0.1	0.14	1.00	1.12	0.23	0.37	1.19	1.25	0.59	0.79	1.29

Find the regression equation of y on x (that is, find the estimated values of C and a).

Solution. Since $y = Ce^{\frac{a}{x}}$

take logarithm on both sides.

$$\ln y = \ln C + \frac{a}{x}$$

Let $z = \ln y$, $u = \frac{1}{x}$, $b = \ln C$

Hence we have

$$z = au + b$$

After substitution with $u = \frac{1}{x}$, $z = \ln y$, the given data is transformed into the data shown in the following table.

u_i	20	16.667	4	3.226	14.286	10	2.632	2.326	7.143	5	2.128
z_i	-2.303	-1.966	0	0.113	-1.470	-0.994	0.174	0.223	-0.528	-0.236	0.255

Using the data (u_i, z_i) and a calculator, we can get

$$|r| = 0.9982833 > 0.602 = r_{0.05} \qquad \text{That is,} \quad |r| > r_{0.05} \text{ linear correlation}$$

We can also get

$$\hat{a} = -0.1459409 \qquad \hat{b} = 0.5476735$$

Hence we have

$$\hat{z} = -0.1459409u + 0.5476735$$

Replacing u and z with the original variables x and y, we have

$$\ln \hat{y} = -\frac{0.1459409}{x} + 0.5476735$$

Therefore

$$\hat{y} = e^{-\frac{0.1459409}{x} + 0.5476735} = e^{0.5476735} \cdot e^{-\frac{0.1459409}{x}} \approx 1.729e^{-\frac{0.1459409}{x}}$$

That is, the regression equation for the curve of y on x is

$$\hat{y} \approx 1.729e^{-\frac{0.1459409}{x}}$$

Exercise 7.2

1. An ice cream shop near a beach sells ice cream at various temperatures, as shown in the following table.

Temperature ($x°$ C)	15.3	16.2	13.9	18.5	21.9	23.1	22.6	25.2
Ice cream sales (y euro)	225	240	198	312	376	422	446	538

a. Calculate $\bar{x}$ and $\bar{y}$, correct to two decimal places.

b. Draw a scatter diagram for the data. Locate the mean point $(\bar{x}, \bar{y})$ on the scatter diagram. Also, draw the line of best fit.

c. Find the equation of the regression line of y on x. Calculate the correlation coefficient r.

d. Describe the relationship between the variables.

e. Estimate the ice cream sales when the temperature is 19°C, correct to two decimal places.

f. Estimate the ice cream sales when the temperature is 28°C, correct to two decimal places.

g. Comment on whether the estimates in e. and f. are reliable.

2. A set of bivariate data is given in the following table.

x	2	4	5	8	12	13	15	19
y	3	5	7	7	16	18	19	21

a. Find the equation of the regression line of y on x. Calculate the correlation coefficient r.

b. Estimate the value of y when $x = 14$, correct to two decimal places.

c. Given that $y = 10$, estimate the value of x, correct to two decimal places.

3. The table below shows students' midterm and final exam scores for a mathematics course. The total marks for the exams are 100.

Midterm exam score x	61	39	70	63	83	75	48	72	54	22	67	60
Final exam score y	83	62	76	77	89	74	48	78	76	51	63	79

a. Construct a scatter diagram.

b. Find the equation of the regression line of y on x.

c. Find the equation of the regression line of x on y.

d. Verify that the mean point $(\overline{x}, \overline{y})$ is located on both regression lines.

e. If a student gets 80 on the midterm exam, what is the student's expected score in the final exam?

f. If a student gets 80 on the final exam, what is the expected score for the student in the midterm exam?

g. A student gets 88 on the midterm exam; explain why the regression equation from b. should not be used to predict the student's score in the final exam.

4. The following table gives the number y of bacteria present in a culture after x hours.

Number of hours (x)	1	2	3	4	5	6	7
Number of bacteria (y)	160	268	476	812	1421	2133	5245

a. Construct a scatter diagram.

b. Find the regression equation of y on x.

c. Estimate the number of bacteria when $x = 4.5$.

5. The following table gives experimental values of the pressure P of a given gas mass corresponding to various values of the volume V. According to thermodynamic principles, a relationship in the form $PV^k = A$, where k and A are constants, should exist between the variables.

Volume V (cm^3)	7.34	7.20	7.14	7.04	6.88	6.68	6.62
Pressure P (Pascal)	100310	103300	103920	105670	107900	111200	112420

a. Find the regression equation of P on V (that is, find the estimated values of k and A).

b. Estimate the value P when $V = 7\ cm^3$.

Answers to exercises

Exercise 1

1. 1320 **2. a.** 17576000 **b.** 15600000 **c.** 12654720 **d.** 11232000

3. 19656 **4. a.** 155117520 **b.** 34767720 **5. a.** 4060 **b.** 3276

6. a. 77520 **b.** 68952 **c.** 50400

Exercise 2.1

1. U = {1, 2, 3, 4, 5, 6, 7, 8, 9, 10, 11, 12, 13, 14, 15, 16, 17, 18, 19, 20}

a. $\frac{1}{2}$ **b.** $\frac{2}{5}$ **c.** $\frac{2}{5}$ **2. a.** $\frac{1}{4}$ **b.** $\frac{7}{12}$ **c.** $\frac{5}{12}$

3. a. U = {BB, BR, BP, RR, RB, RP, PP, PB, PR} **b.** $\frac{2}{3}$

4. U = {HHH, HHT, HTH, THH, HTT, THT, TTH, TTT}

a. $\frac{1}{8}$ **b.** $\frac{3}{8}$ **c.** $\frac{3}{8}$ **d.** $\frac{1}{8}$ **e.** $\frac{7}{8}$ **f.** $\frac{1}{2}$

5. a. U = {(5, 6), (5, 7), (5, 8), (6, 5), (6, 7), (6, 8), (7, 5), (7, 6), (7, 8), (8, 5), (8, 6), (8, 7)} **b.** $\frac{1}{3}$ **6. a.** $\frac{3}{14}$ **b.** $\frac{27}{56}$ **c.** $\frac{27}{28}$ **d.** $\frac{15}{56}$

7. U = {BBB, BBG, BGB, BGG, GBB, GBG, GGB, GGG}

a. $\frac{3}{8}$ **b.** $\frac{1}{2}$ **c.** $\frac{1}{2}$ **d.** $\frac{3}{8}$ **e.** $\frac{1}{2}$ **8. a.** $\frac{7}{18}$ **b.** $\frac{7}{36}$ **c.** $\frac{11}{18}$ **9.** $\frac{2}{7}$

10. a. $\frac{1}{17}$ **b.** $\frac{13}{51}$ **c.** $\frac{25}{51}$ **11. a.** $\frac{28}{45}$ **b.** $\frac{1}{45}$ **c.** $\frac{16}{45}$ **d.** $\frac{1}{5}$

12. a. 0.504 **13. a.** $\frac{1}{12}$ **b.** $\frac{1}{20}$ **14.** $\frac{41}{81}$ **15. a.** $\frac{1}{9979200}$ **b.** $\frac{1}{415800}$

Exercise 2.2

1. a. $\frac{1}{6}$ b. $\frac{91}{216}$ 2. a. 0.4 b. 0.6 c. 0.1

3. a. 0.9 b. 0.4 c. 0.1 4. a. 0.5 b. 0.7 c. 0.1

5. a. 0.8 b. 0.55 c. 0.75 d. 0.2 6. a. $\frac{1}{3}$ b. $\frac{2}{3}$ c. $\frac{1}{12}$

7. a. 0.7 b. 0.5 8. a. $\frac{776}{825}$ b. $\frac{1}{1650}$ c. $\frac{49}{825}$ 9. $\frac{5}{6}$

10. 0.5 11. a. 0.25 b. 0.35 c. 0.5 d. 0.5 12. $\frac{mN+n(N+1)}{(n+m)(N+M+1)}$

Exercise 2.3

1. $\frac{1}{3}$ 2. a. $\frac{1}{2}$ b. $\frac{1}{2}$ c. 0 d. $\frac{1}{3}$ 3. $\frac{1}{2}$

4. a. $\frac{15}{91}$ b. $\frac{15}{91}$ c. $\frac{5}{21}$ d. $\frac{9}{13}$ 5. a. $\frac{5}{12}$ b. $\frac{2}{5}$ 6. $\frac{1}{3}$

7. A and B are not independent. 8. 0.4 9. a. $\frac{1}{3}$ b. $\frac{1}{3}$ c. $\frac{2}{3}$

d. $\frac{1}{2}$ 10. 0.6928 11. A and B are independent. 12. a. 0.4 b. $\frac{690}{1421} \approx 0.486$

13. a. $\frac{1}{12}$ b. $\frac{1}{9}$ c. $\frac{1}{3}$ 14. a. $\frac{3}{4}$ b. $\frac{1}{2}$ c. $\frac{7}{12}$ d. $\frac{5}{8}$ e. $\frac{3}{8}$ 15. a. $\frac{1}{2}$

b. $\frac{5}{12}$ c. $\frac{3}{5}$ 16. a. $\frac{1}{12}$ b. $\frac{1}{2}$ c. $\frac{5}{12}$ d. $\frac{11}{12}$ e. 16 17. $\frac{5}{8}$

Exercise 2.4

1. a. $\frac{21}{55}$ b. $\frac{27}{55}$ c. $\frac{34}{55}$ d. $\frac{3}{44}$ e. $\frac{3}{11}$ 2. a. $\frac{1}{5}$ b. $\frac{2}{5}$ c. 1

d. 0 3. $\frac{19}{100}$ 4. 0.3 5. $\frac{3}{2n-1}$ 6. 0.179 7. $\frac{1}{55}$ 8. $\frac{5}{9}$

9. **a.** $\frac{29}{90}$ **b.** $\frac{20}{61}$ **10.** 0.00344 **11.** $\frac{3}{17}$ **12.** 0.7

13. **a.** $\frac{1}{18}$ **b.** $\frac{19}{36}$ **14.** $\frac{5}{11}$ **15.** 0.957 **16.** 8.27×10^{-8}

17. **a.** $p_A p_B p_C$ **b.** $1-(1-p_A)(1-p_B)(1-p_C)$

c. $1-(1-p_A^2)(1-p_B^2)(1-p_C^2)$ **d.** $p_D^2[1-(1-p_A)(1-p_B)(1-p_C)]$

Exercise 2.5

1. **a.** 0.8992 **b.** $\frac{138}{281}$ **2.** $\frac{294}{1291}$ **3.** It is more likely from the first box. **4.** $\frac{35}{68}$ **5.** $\frac{3}{13}$ **6.** $\frac{196}{197}$ **7.** $\frac{4}{9}$ **8.** **a.** $\frac{7}{600}$ **b.** $\frac{3}{7}$

9. $\frac{98}{193}$ **10.** **a.** $\frac{1}{2}$ **b.** from the second pen holder **11.** $\frac{20}{141}$

12. **a.** $\frac{27}{49}$ **b.** $\frac{2}{9}$ **13.** $\frac{4}{19}$ **14.** **a.** $\frac{20}{39}$ **b.** $\frac{34}{55}$ **15.** $\frac{15}{38}$

Exercise 3.1

1. $\frac{1}{4}$ **2.** **a.** $\frac{1}{20}$ **b.** $\frac{3}{20}$ **c.** 0.75

3.

X	3	4	5
p	0.1	0.3	0.6

4. $\frac{60}{77}$

5. a.

X	0	1	2
p	$\frac{68}{95}$	$\frac{51}{190}$	$\frac{3}{190}$

b.

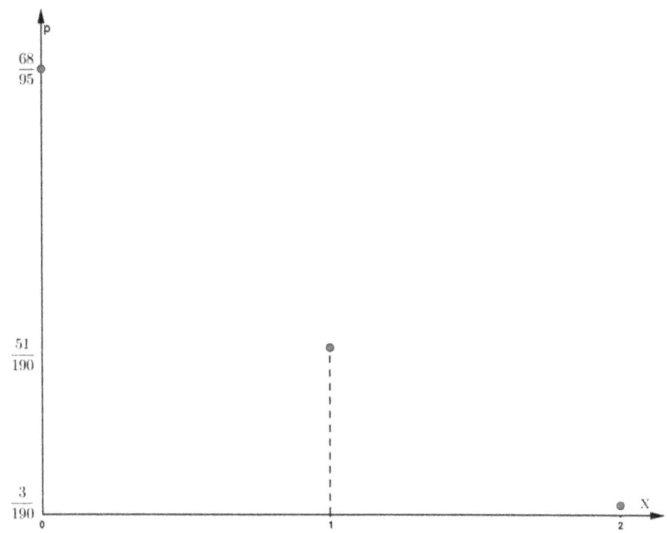

6. $\frac{1}{5}$

7.

X	1	2	3	4	5
p	0.1	0.09	0.081	0.0729	0.06561

Exercise 3.2

1. a. 0.140 **b.** 0.0424 **c.** 0.00633 **2.** $\frac{2}{3}$ **3. a.** 0.4096

b. 0.73728 **4. a.** 0.65625 **b.** 3 **5. a.** $1-(1-p)^m$ **b.** $\binom{m}{10}p^{10}(1-p)^{m-10}$

c. $1-(1-p)^m-\binom{m}{1}p(1-p)^{m-1}-\binom{m}{2}p^2(1-p)^{m-2}$ **6. a.** 0.268 **b.** 0.522

7. a. 0.201 **b.** 0.633 **8.** 9 **9.** 0.607

10. It is better to invite 5 representatives. **11.** 0.259 **12.** 0.180 **13.** 0.479

14. e^{-1} **15.** $n \geq 16$ **16.** 0.933 **17.** $\frac{241}{969}$ **18.** $\frac{23}{102}$

Exercise 3.3

1. a.

$$F(x)=\begin{cases} 0 & x<-2 \\ \dfrac{1}{5} & -2\le x<1 \\ \dfrac{7}{10} & 1\le x<3 \\ 1 & x\ge 3 \end{cases}$$

b. **i.** $\frac{1}{5}$ **ii.** $\frac{1}{2}$ **iii.** $\frac{3}{10}$

2.

X	-3	-2	3
p	0.3	0.2	0.5

3. a. no **b.** yes **c.** yes **d.** yes **e.** no **4. a.** k = 6 **b.** 0.896

5. a. **c** = $\frac{1}{9}$ **b. i.** $\frac{1}{2}$ **ii.** $\frac{3}{4}$ **6.** k ≈ 0.613 **7.** $a=\frac{1}{3}, b=\frac{1}{2}$

8. a. $\frac{1}{\pi}$ **b.** $\frac{1}{2}$ **9. a.** a = 1 **b.** $\frac{3}{4}$

Exercise 3.4

1. a. 0.189 **b.** 0.0883 **c.** 0.487 **d.** 0.632 **2. a.** 0.368 **b.** 0.368

3. 0.368 **4.** $\frac{1}{2}$ **5. a.** 0.252 **b.** 0.539 **c.** 0.252 **d.** 0.516

6. 4305 **7. a.** 0.921 **b.** 32.1 **c.** 36.9 **8.** 72.2%

9. a. 0.977 **b.** 0.954 **c.** a > 1.64 **10. a.** 0.625, 0.267 **b.** 0.139

11. a. 0.8 **b.** 0.954 **12.** 15.6 **13.** $\mu \approx 3.88$, $\sigma \approx 0.199$

14. a. 0.0608 **b.** 732 **c. i.** 0.0134 **ii.** 0.999 **15. a.** 1.96

b. 1.64 **16.** $\mu \approx 67.9$, $\sigma \approx 1.50$ **17. a.** 771.88 **b.** 0.02599 **c.** 0.0943

18. a. i. 0 **ii.** 0.0478 **b.** 37 < X < 53 **c.** 0.152

19. a. $\mu \approx 60.7$, $\sigma \approx 23.9$ **b.** 0.969 **20. a.** 0.4 **b.** 0.913

Exercise 3.5

1.

Y	1	0	2
p	0.3	0.4	0.3

2.

$$f_Y(y)=\begin{cases}\dfrac{1}{y} & 1<y<e\\ 0 & elsewhere\end{cases}$$

3. a. $f_Y(y)=\dfrac{1}{\pi(1+y^2)} \qquad -\infty<y<\infty$

b. $f_Z{}^*(z)=\begin{cases}\dfrac{2}{\pi\sqrt{1-z^2}} & 0<z<1\\ 0 & elsewhere\end{cases}$

Exercise 3.6

1. a.

X	$n \ (n\in Z^+)$
p	$0.6^{n-1}\cdot 0.4$

b. 0.375 **c.** 0.36

2. a.

X	0	1	2	3	4
p	0.2401	0.4116	0.2646	0.0756	0.0081

b. 0.59526 **c.** 0.415 **3.** 0.635 **4.** $\frac{1}{36}$ **5.** $\frac{9}{19}$ **6.** 0.936

7. $p_4=0.1296$, $p_5=0.20736$, $p_6=0.20736$, $p_7=0.165888$ **8.** 0.6

9. a. 0.677 **b.** 0.238 **c.** 0.458 **10.** $\frac{20}{27}$ **11. a.** 702.83 **b.** 0.880

Exercise 4

1. a. 0.8 **b.** 1.8 **c.** 1.16 **d.** 3.8 **e.** 74.24 **f. i.** 4.4 **ii.** 11.44

2. a. -0.5 **b.** 2.25 **c.** -3.5 **d.** 2.25 **e. i.** $\frac{7}{12}$ **ii.** $\frac{17}{144}$

3. $n = 16, p = 0.625$ **4. a.** $\frac{2}{9}$ **b.** 2 **c.** 0.5 **d.** E(Y) = -6, Var(Y) = 8

5. a. $\frac{\sqrt{30}}{2}$ **b.** $\frac{8}{3}$ **c.** $\frac{\sqrt{38}}{6}$ **6.** a = 12, b = -12, c = 3 **7.** $E(X) = \frac{1}{6}$, $Var(X) = \frac{5}{36}$

8. $E(X) = 75$, $Var(X) = 74.775$ **9.** $E(X) = 0.87$, $Var(X) = 0.5781$

10. a. n = 223 **b.** $E(X) = 0.223$, $sd(X) \approx 0.472$ **c.** 7.787 **11.** 0.0571

12. a. 0.08 **b.** 16 **13. a.** 60 **b.** 5.4 **14.** 1110 g **15.** 2 **16.** -4.29

17. a.

X	0	1	2
p	$\frac{11}{42}$	$\frac{101}{168}$	$\frac{23}{168}$

b. 175

18. $E(X) = 4.5$, $Var(X) = 0.45$ **19.** $\frac{8}{9}$ **20.** $\frac{1}{37}$

Exercise 5

1. 0.132 **2. a.** 0.0143 **b.** 449 **3.** 0.913 **4.** 0.0950 **5.** 0.0127 **6.** 0.309

Exercise 6

1. a. $\overline{x}_A = 100$, $\overline{x}_B = \frac{690}{7} \approx 98.57$, $s_A^2 = \frac{24}{7} \approx 3.43$, $s_B^2 = \frac{9700}{49} \approx 198$

b. Machine A **2.** $s_A^2 = 0.00954$, $s_B^2 = 0.02434$ Athlete A has more stable performance. **3.** $n = 10$ **5. a.** All are unbiased estimators of μ. **b.** Y_1 has the smallest variance.

Exercise 7.1

1. There is a relationship. **2.** There is a relationship. **3.** There is a relationship.

4. There is a difference. **5.** The effect is significant.

Exercise 7.2

1. a. $\overline{x} \approx 19.59$, $\overline{y} \approx 344.63$

b.

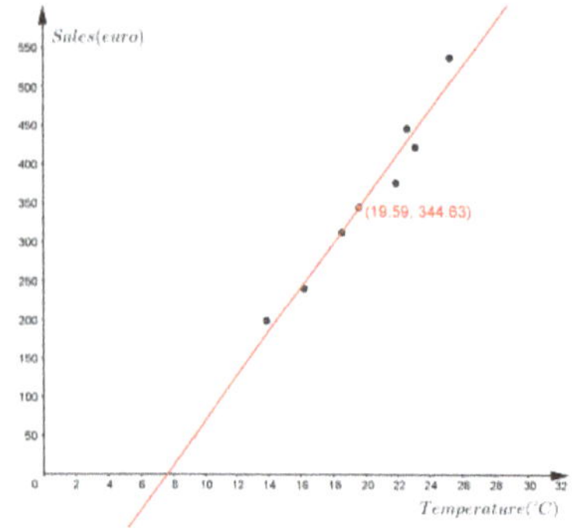

c. $\hat{y} = 28.5392369x - 214.3873$, r = 0.98336564

d. very strong positive linear correlation **e.** 327.86 **f.** 584.71

g. The estimate in e. is an interpolation, and there is a very strong positive linear correlation between the variables. We therefore expect this estimate to be reliable. The estimate in f. is an extrapolation, so it may not be reliable.

2. a. $\hat{y} = 1.17575757x + 0.53636363$, r = 0.97218989 **b.** 17.00 **c.** 8.14

3. a.

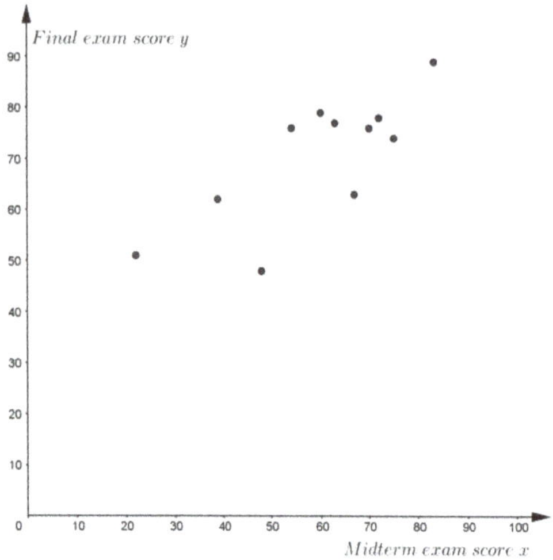

b. $\hat{y} = 0.56556588x + 37.682163$ **c.** $\hat{x} = 1.00876858y - 12.458825$

d. $(\overline{x}, \overline{y}) = (59.5, 71.3333333)$ locates on both regression lines.

e. 82.9274334 **f.** 68.2426614

g. The estimate is an extrapolation so the estimate may not be reliable.

4. a.

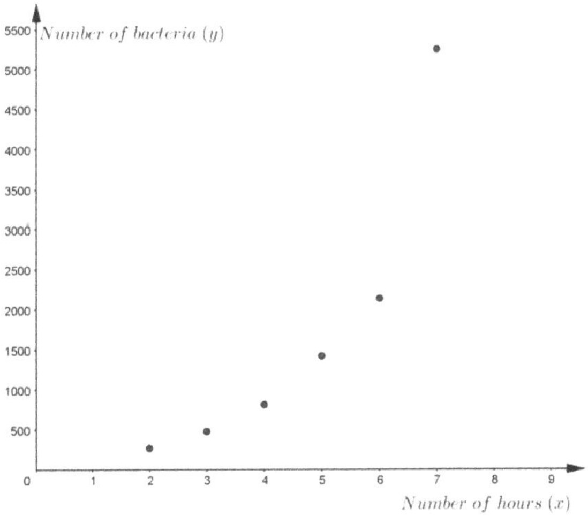

b. $\hat{y} = 87.5897e^{0.56113799x}$ **c.** 1094.206111

5. a. $\hat{P}V^{1.0563016} = 827994.0936$ **b.** 106010.4864

www.ingramcontent.com/pod-product-compliance
Lightning Source LLC
LaVergne TN
LVHW060029210926
848481LV00013B/3823